Superhydrophobic Coating - Recent Advances in Theory and Applications

Edited by Junfei Ou

Published in London, United Kingdom

Superhydrophobic Coating - Recent Advances in Theory and Applications
http://dx.doi.org/10.5772/intechopen.111048
Edited by Junfei Ou

Contributors
Brenda Resendiz Diaz, Colin R. Crick, Zichen Zhang, Junfei Ou, Wen Li, Syafiqa M. Saleh, Athirah Mohd Tamidi, Farahdila Kadirkhan, Pei Ching Oh, Reyna I. Garcia-Gonzalez, Mohamed S. Selim, Hamed I. Hamouda, Nesreen A. Fatthallah, Mohsen S. Mostafae, Shimaa A. Higazy, Samah Shabana, Ashraf M. EL-Saeed, Zhifeng Hao, Katarzyna Buczkowska

First published in London, United Kingdom, 2024 by IntechOpen
IntechOpen is the global imprint of INTECHOPEN LIMITED, registered in England and Wales, registration number: 11086078, 5 Princes Gate Court, London, SW7 2QJ, United Kingdom

British Library Cataloguing-in-Publication Data
A catalogue record for this book is available from the British Library

Additional hard and PDF copies can be obtained from orders@intechopen.com

Superhydrophobic Coating - Recent Advances in Theory and Applications
Edited by Junfei Ou
p. cm.
Print ISBN 978-0-85466-281-4
Online ISBN 978-0-85466-280-7
eBook (PDF) ISBN 978-0-85466-282-1

For EU product safety concerns:
IN TECH d.o.o., Prolaz Marije Krucifikse Kozulić 3, 51000 Rijeka, Croatia,
info@intechopen.com or visit our website at intechopen.com.

Meet the editor

Junfei Ou earned his Ph.D. in Surface Engineering from the Lanzhou Institute of Chemical Physics, Chinese Academy of Science, in 2010. Following this, he joined Nanchang Hangkong University, China, as an assistant professor and was promoted to associate professor in 2014. In 2018, he transitioned to the Jiangsu University of Technology (JSUT), China, where he became a full professor in 2020. Currently, he serves as the dean of the School of Materials Engineering at JSUT. Dr. Ou's research focuses on surface and interface studies, notably in the field of bio-inspired superhydrophobic coatings. He has authored more than 130 SCI papers on surface and interface topics.

Contents

Preface **IX**

Chapter 1 **1**
Superhydrophobic Coatings with Environmentally Friendly Materials
by Brenda Resendiz Diaz and Colin R. Crick

Chapter 2 **23**
Perspective Chapter: Challenges in the Durability of Superhydrophobic Coatings Pertinent to Unmanned Aerial Vehicle (UAV) Icing Mitigation
by Zichen Zhang, Junfei Ou and Wen Li

Chapter 3 **49**
Superhydrophobic Membrane for Gas-Liquid Membrane Contactor Applications
by Syafiqa M. Saleh, Athirah Mohd Tamidi, Farahdila Kadirkhan and Pei Ching Oh

Chapter 4 **71**
The Effect of Micro/Nano Roughness on Antifouling and Bactericidal Surfaces
by Reyna I. Garcia-Gonzalez and Colin R. Crick

Chapter 5 **91**
Advanced Bioinspired Superhydrophobic Marine Antifouling Coatings
by Mohamed S. Selim, Hamed I. Hamouda, Nesreen A. Fatthallah, Mohsen S. Mostafae, Shimaa A. Higazy, Samah Shabana, Ashraf M. EL-Saeed and Zhifeng Hao

Chapter 6 **121**
Hydrophobic Protection for Building Materials
by Katarzyna Buczkowska

Preface

Bio-inspired superhydrophobic coating, which exhibits exceptional nonwetting properties characterized by a large contact angle and minimal contact angle hysteresis, has garnered significant attention across a broad range of applications. This book comprehensively and systematically explores the various advancements pertaining to superhydrophobic coatings, starting from their fabrication and extending to their practical implementation. The book delves into eco-friendly methods of producing superhydrophobic coatings, highlighting the importance of sustainable approaches. Furthermore, it explores the notable application scenarios wherein superhydrophobic coatings have proven to be effective, including anti-icing, anti-fouling, and gas–liquid separation.

Chapter 1 provides a comprehensive overview that delves into the latest techniques employed in the production of superhydrophobic coatings. These techniques are specifically focused on utilizing environmentally safe materials, thereby addressing growing concerns regarding their impact on the environment.

Chapter 2 reviews recent advancements in durability tests specifically focused on superhydrophobic coatings that are applicable to anti-/de-icing in unmanned aerial vehicles. Additionally, this chapter sheds light on the often-neglected aspect of superhydrophobic coating degradation caused by droplet erosion, which can result in significant harm to the surface topology due to the application of "water hammer pressure." Consequently, this study offers valuable recommendations to promote a more systematic and efficient approach toward conducting durability tests on superhydrophobic coatings.

Chapter 3 explores the extensive research conducted on gas–liquid membrane contactors. Here, the primary emphasis is placed on superhydrophobic membranes that have been ingeniously designed to tackle the ever-present challenge of membrane pore wetting. By addressing this issue, researchers aim to optimize the performance and efficiency of these contactors.

Chapter 4 takes a deep dive into the intriguing realm of micro/nano roughness and its profound influence on the development of antifouling and bactericidal surfaces. Through this exploration, the chapter seeks to enhance our understanding of how these surfaces effectively combat microorganisms, ultimately offering improved strategies for successful coexistence between humans and microorganisms.

Chapter 5 elucidates the evolutionary journey undertaken by superhydrophobic antifouling surfaces that have been ingeniously inspired by biological systems. Additionally, this chapter introduces an innovative and state-of-the-art research platform aimed at propelling the development of next-generation antifouling surfaces specifically tailored for maritime navigation, thereby revolutionizing the field.

Chapter 6 highlights the significance of modifying wetting properties to augment the durability and performance of construction materials. This chapter emphasizes the pivotal role played by nanomaterials and bulk additives in the transformative alteration of surface properties. Furthermore, investigations into coatings specifically formulated to provide surface protection are meticulously outlined. The practical applications of hydrophobic coatings are exemplified through a diverse array of commercial products, effectively illustrating their real-world relevance and significance.

By addressing these topics, the book provides readers with a comprehensive understanding of the historical background and the remarkable progress made in the field of superhydrophobic coatings. Moreover, this book serves as a rallying call to scientists worldwide, urging them to collectively overcome the barriers impeding the development of lotus-inspired superhydrophobic coatings and pave the way for their exploration in new and untapped application domains.

Junfei Ou
School of Materials Engineering,
Jiangsu University of Technology,
Changzhou, China

Chapter 1

Superhydrophobic Coatings with Environmentally Friendly Materials

Brenda Resendiz Diaz and Colin R. Crick

Abstract

Due to their excellent robustness and water-repellence properties, materials with low surface energy such as fluoroalkyl compounds (perfluoroalkyl silanes and fluoroacrylic copolymers) and organosilane-based chemistries are used for superhydrophobic coatings fabrication. However, these materials can cause a severe environmental impact and generally are not biodegradable or recyclable. For this reason, new environmentally friendly methods using natural materials are still being developed to obtain similar features, especially for packaging, textile and medical applications. The use of plant-based materials shows potential for creating superhydrophobic coatings, as many of them are naturally hydrophobic and can produce the desired surface textures. The main challenges to making superhydrophobic coatings from plant-based materials are abrasion resistance, strong adhesion, functionality in certain environments, and durability, but at the same time, they must be biodegradable. This chapter summarizes the recent approaches for superhydrophobic coatings made from environmentally safe materials and their applications.

Keywords: superhydrophobic coatings, plant-based materials, eco-friendly, biopolymers, sustainable

1. Introduction

In nature, superhydrophobicity is observed in various plants, animals, and insects. They possess a unique surface texture that allows them to repel water effortlessly. Researchers are focusing on mimicking such surface structures for real-world applications to generate properties such as surface protection, anticorrosion, anti-icing, self-cleaning, antifouling, and so on [1]. To fabricate an artificial superhydrophobic surface, two main factors must be considered (1) hierarchical micro- and/or nanoscale roughness and (2) surface chemistry [2]. Due to their low surface energy, synthetic materials are typically used for superhydrophobic coatings fabrication [3]. Superhydrophobic coatings that have been traditionally used are made from fluorinated and sulfhydryl compounds, along with silicones. However, these materials are harmful to the environment, toxic, and nondegradable, and they are relatively expensive. Moreover, their fabrication involves the use of organic solvents, which makes them challenging to dispose of and keep human health safe [4, 5].

Despite this, they successfully achieve lower surface free energy and provide a degree of durability [4]. Recently, researchers have been placing importance on substituting these materials as their main priority. New eco-friendly alternatives including bio-based polymers such as cellulose, chitosan, and plant-based waxes are being implemented to produce robust coatings [4–8]. The advantages of these materials are their biodegradability, non-toxicity, and low-cost, which makes them good prospects for industries like food packaging [4]. However, most plant-based materials are hydrophilic, which limits their application. As a result, chemical modifications and new innovative techniques are necessary to achieve properties such as hydrophobicity, resilience, and longevity [4, 7, 8].

When fabricating superhydrophobic coatings, materials such as cellulose, chitosan, and lignin are commonly used as rough agents, while waxes and fatty acids are utilized as low surface energy materials [8]. The combination between them leads to obtaining contact angles above 150°, which is the value to consider a surface superhydrophobic [9]. This chapter aims to provide a summary of common biopolymers and other plant-derived materials used for fabricating superhydrophobic coatings, along with the modifications needed to achieve hydrophobicity properties considering biosafety and cost-effective scale-up.

2. Biopolymers

2.1 Definition and classification

Polymers that are derived from renewable sources, carbon neutral, and can biodegrade are known as bio-based polymers. This encompasses all plants, animals, or microorganisms' mass resources [10, 11]. Different biopolymer classifications depend on the production process, the source material, and the material's life cycle [11]. They can be classified into three categories based on their origin and production (**Figure 1**) [14]:

1. Polymers extracted from biomass, that is, polysaccharides such as starch, cellulose, and proteins.

2. Polymers made by conventional chemical synthesis using bio-based monomers, that is, polylactic acid.

3. Polymers produced by microorganisms or genetically modify them through bacteria, that is, polyhydroxyalkanoate.

Biopolymers as raw materials are gaining importance in various industries. These polymers are preferred by companies due to their environmentally friendly and non-toxic nature [13]. Sectors such as food packaging, textiles, cosmetics, and medicine are working toward achieving similar properties as synthetic polymers [15].

Biopolymers are a potential alternative to film fabrication due to their biodegradability, renewability, and low-cost [10]. Many of these substances exhibit hydrophilic behavior (presence of -OH groups), resulting in reduced resilience, poor cohesion and adhesion, and inferior mechanical properties. To expand their application fields, it is necessary to modify them to make them hydrophobic chemically [10].

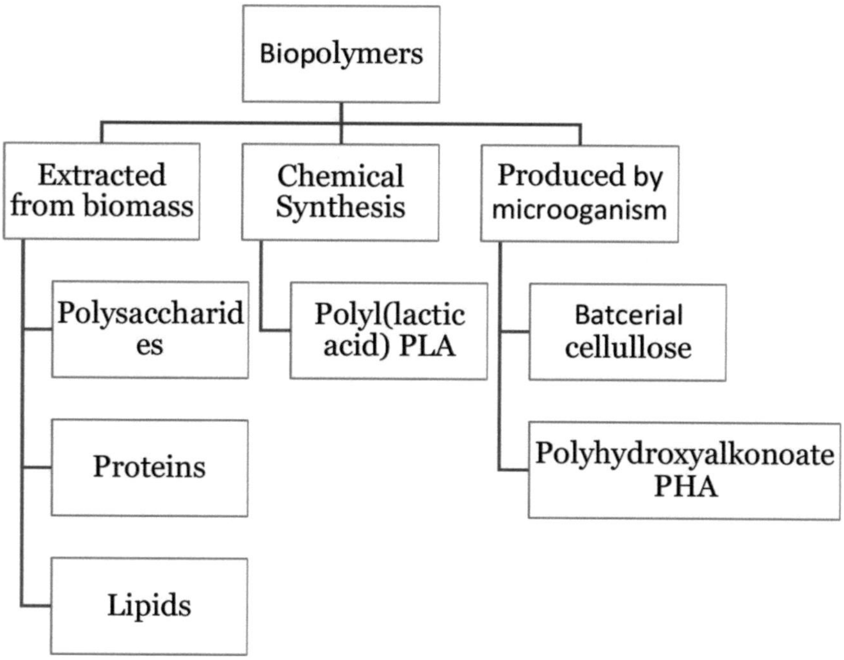

Figure 1.
Biopolymers classification [12, 13].

3. Superhydrophobic coatings from polysaccharides

3.1 Polysaccharides

Polysaccharides are created from monosaccharides that can be found in plants such as cellulose, alginates, and starch, as well as in animals like chitosan [16, 17]. They are commonly used as a raw material; however, the presence of hydrophilic groups leads them to have poor water repellence, which limited them for certain applications [10]. Researchers are developing new strategies to enhance hydrophobic behavior to fabricate novel superhydrophobic surfaces from polysaccharides that mainly work as roughness agents [4, 7, 8].

3.1.1 Cellulose

Cellulose is considered one of the most eco-friendly, renewable, and biodegradable materials in the world [15]. Primarily, cellulose is extracted from the cell wall of woody plants and is combined with lignin and hemicellulose [18]. It has been established that cellulose is composed of glucose units arranged in a polymer structure (poly-b-(1,4)-D-glucose) and hydroxyl groups in an equatorial position [15, 19, 20]. The material's strength is attributed to the numerous hydroxyl groups that form strong hydrogen bonds [18].

Chemical modification of cellulose is required to replace hydroxyl groups with hydrophobic groups to achieve superhydrophobic properties [21, 22]. A variety of strategies have been performed to enhance the efficiency of the surface hydrophobicity to improve compatibility and dispersibility in different solvents [23]. Recently, cellulose has been used as a coating ingredient to generate surface roughness by

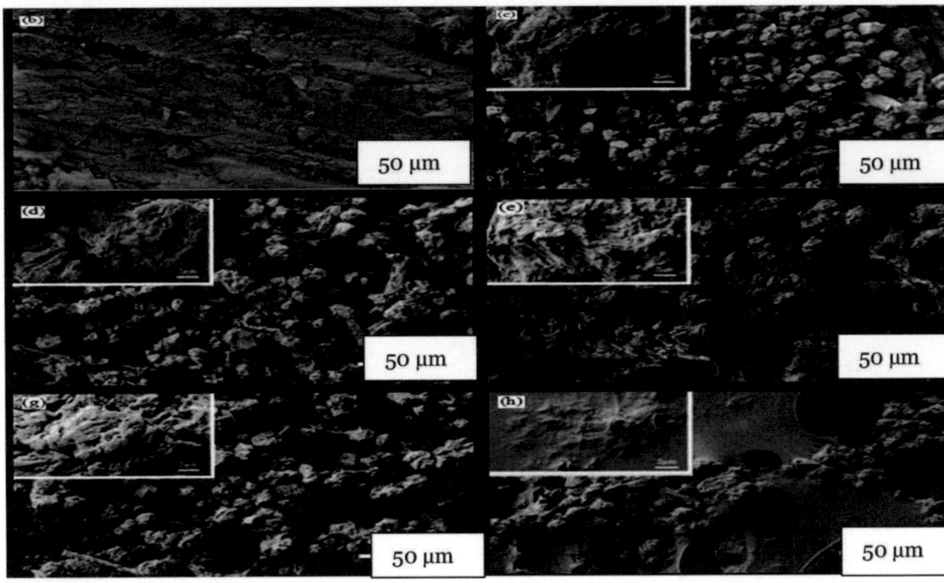

Figure 2.
Surface morphology of L-CNC coating at higher magnification and its particle surface. The coating presented high roughness. Reproduced (adapted) from Ref. [24] with permission from the Royal Society of Chemistry.

forming nano/micro cellulose fibrils or cellulose nanoparticles [8]. To avoid using hazardous organic solvents, researchers developed lignin-coated cellulose nanoparticles (L-CNC). Lignin which is less hydrophilic than cellulose, was used to build a rough structure which was then combined with a water-soluble nontoxic polymer, polyvinyl alcohol (PVA) which acts as a binder material (**Figure 2**) [24]. Using the spray coating technique, lignin-cellulose nanoparticles/PVA solutions demonstrated superhydrophobicity with water contact angles around 150° [25].

A green and effective nanofibrillated cellulose-based superhydrophobic coating was built by using a coupling reaction of ethyl orthosilicate and cetyltrimethoxysilane. The coating revealed nano roughness, low surface energy, and good thermal stability. Additionally, the coating exhibited a certain degree of antibacterial activity. The wettability of the surface was monitored for 30 days, and it demonstrated good water resistance (contact angle above 160°) [26]. A semitransparent superhydrophobic surface with a rough texture was created using nano- and microstructure cellulose by utilizing the spray coating technique. To make this surface, building blocks were prepared and then molded to mimic the surface of lotus leaves. After substituting the hydroxyl groups of cellulose with stearoyl groups, the static contact angle was significantly increased to over 158°, and the contact angle hysteresis was lowered to 5°. As a result, it exhibited excellent water repellency against a jet of water impinging [27].

3.1.2 Chitin and chitosan

Chitin is the second most abundant biopolymer in nature, which is derived from the exoskeleton of crustaceous as well as from the fungi and insects' cell walls [28]. It is a polymer with a long chain of β-(1,4)-N-acetyl-D-glucosamine and N-acetyl-D-glucosamine [15, 18]. Chitosan is derived from the N-deacetylation of chitin using alkali [15, 28]. The degree of deacetylation varies from 40 to 98%, which determines the molecular weight of the polymer and its applications [28]. Chitosan is a weak base and insoluble in water, which is commonly used for medical and pharmaceutical applications [15].

Chitosan micro/nanoparticles have the potential to be used in superhydrophobic coatings [8]. The physicochemical properties of these particles, such as crystallinity, molecular weight, and functional groups, can vary depending on the method used to create them. The particles can have either a spherical or an amorphous morphology. Chitosan, due to its amino and hydroxyl functionality, is suitable for chemical modification, such as deacetylation [23].

Chitosan-based nanoparticles offer certain advantages for medical textiles such as antibacterial activity and low toxicity [29]. A proposed superhydrophobic coating using chitosan nanoparticles was made through the covalent bonding of octadecylamine (ODA) to chitosan with the help of a cross-linking agent, glutaraldehyde (GA) (**Figure 3**). To reduce surface energy, ODA has been used thanks to its long hydrocarbon chain, while particle accumulation (measuring 200 nm in size) was used to enhance roughness. The process involved the extraction of chitin from crab shells, followed by cleaning and grinding, decolorization, deproteinization, demineralization,

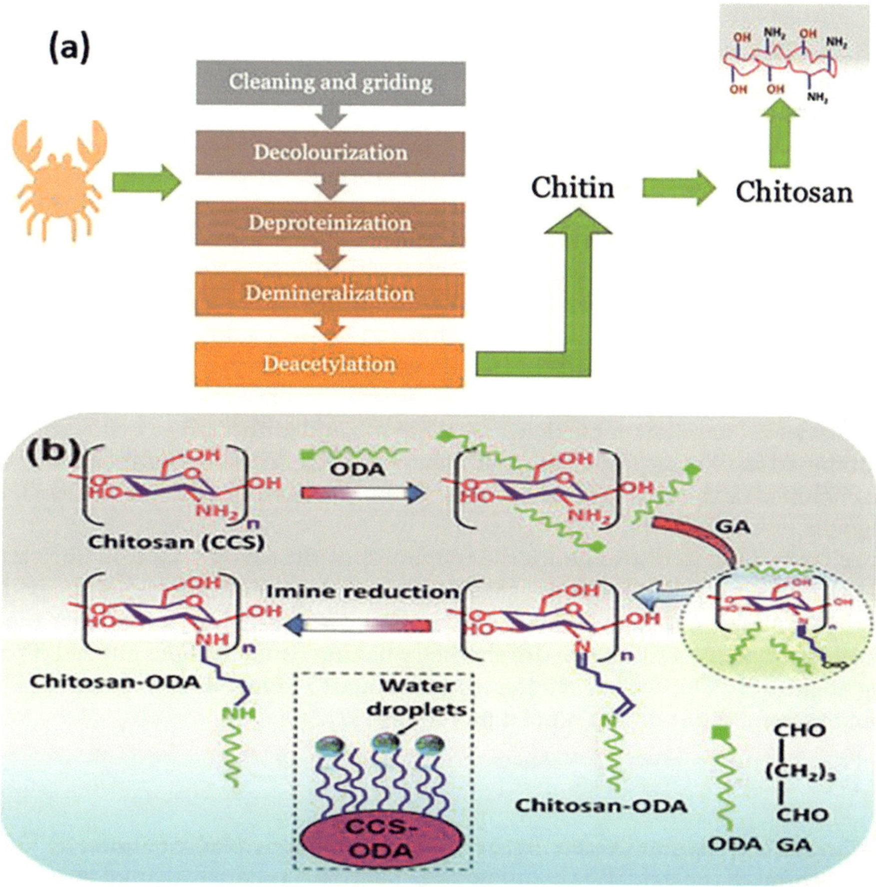

Figure 3.
Crab shells are processed to obtain chitosan powder for the fabrication of chitosan-based superhydrophobic material. Reprinted (adapted) with permission from [30] Copyright 2023 American Chemical Society.

and deacetylation. Afterward, chitosan underwent chemical modification with GA and ODA before being sprayed onto polyester fabric, medical-grade cotton, and PU sponge using the spray coating technique, obtaining water contact angles for all materials above 158° [30].

A safe and environmentally sustainable fabrication method was used to obtain a chitosan-fluorine-free superhydrophobic coating to be transparent. The process consists of two main steps: (1) chitosan functionality modification and (2) solvent-free deposition. During the first stage, stearoyl chloride reacts with the hydroxyl and amine groups on the chitosan backbone, resulting in esterification and secondary amide formation. In the second stage, the modified chitosan is applied onto a glass slide and heated in an oven to ensure it adheres to the substrate. It was demonstrated that by increasing the number of chitosan, the hydrophobicity enhances, leading to higher contact angles (>150°) [31].

3.1.3 Starch

Plants such as corn, rice, and potato contain abundant biopolymer known as starch, which can be stored in various parts of the plant like roots, tubers, leaves, and seeds [15, 32]. It is insoluble in water, semicrystalline, and dense [32]. These carbohydrates consist of linear polysaccharide amylose and branched polysaccharide, which are formed at the end of photosynthesis [15]. It is composed of glucose units linked to α-1,4 and α-1,6 glycosidic bonds. Starch is a highly adaptable biomaterial that has garnered attention for its widespread availability, lack of toxicity, affordability, and ability to biodegrade [33].

The high performance of nano-based superhydrophobic coatings has been increasing interest in a variety of sectors [34, 35]. Starch nanoparticles show excellent micro/nano structure to form environmentally friendly superhydrophobic coatings. By combining a lower surface tension material (i.e., PDMS) covering and nano starch-based particles, the robustness water repellent coatings can be achieved with a water contact angle >150° [35]. It is important to notice that pH could help to monitor the freshness of the coating when it is in contact with different types of food [34]. A thermostable and colorimetric starch-based superhydrophobic coating (>150°) with edible materials (starch nanoparticles, stearic acid, and anthocyanin) was fabricated. The combination of stearic acid and starch nanoparticles contributes to the micro/nano roughness of the coating, while anthocyanin is used to monitor freshness levels through pH levels [36].

Starch is mainly used to enhance the roughness of the surface. By utilizing starch-based materials, polyethyleneimine (utilized as an interlayer binder), and beeswax, a superhydrophobic coating was created. Beeswax density of 1.1 mg cm^{-2} was the correct value to maintain superhydrophobicity having contact angles above 150° and a sliding angle of 6°. The coatings indicated resistance to immersion of water for a long time and showed the ability to repel liquid foods [37].

3.1.4 Lignin

Lignin is a biopolymer present in the cell walls of woody plants (**Figure 2**) [38]. It is formed through the oxidative coupling of p-hydroxy cinnamyl alcohol monomers and other related compounds. The cell wall gains rigidity, strength, and increased resistance through the covalent bonding of lignin and hemicellulose [39, 40].
The structural organization of lignin contributes to the physical and mechanical

properties of wood [41]. It is a natural and renewable material that has recently been exploited as raw material for industrial applications [38]. However, chemical modification is necessary to enhance hydrophobicity due to its hydrophilicity and poor dispersibility [42].

Lignin-based superhydrophobic coatings are still developed due to the complex chemical structure and difficult modification of lignin [8]. Recently, lignin nanospheres have been successfully fabricated for electrochemistry utilization to form rough structures, fluorine-free silane coupling to reduce the surface energy, and nanocellulose crystals to act as reinforcement material. The hydrothermal lignin nanospheres without modification reached a contact angle of 139° for just 1 minute after the surface become hydrophilic. However, by combining both methyltrimethoxysilane and hexadecyltrimethoxysilane, the coating reached a contact angle of 164° [43]. A different instance of utilizing lignin for constructing a superhydrophobic coating involved coating cellulose nanocrystals with lignin. Varied particle sizes aid in achieving an appropriate rough surface structure. To create the coating, the lignin particles were first modified to have low surface energy. Then, they were placed on the substrate that had been covered with adhesive. Finally, a wooden bar was used to press the particles into the adhesive, resulting in contact angles above 160°(**Figure 4**) [25].

Lately, a green effective approach using lignin-micro-nanospheres (LMNS) chemically modified with γ-Valerolactone (GVL) was proposed. The lignin microspheres were utilized for the fabrication of superhydrophobic coatings on wood surfaces by immersing them in the solution (LMN, ethanol, F13-TMS, and epoxy). It was demonstrated that LMNS were uniformly anchored to the wood surface and showed spherical morphology. The hierarchical micro/nano structure provides the surface with hydrophobicity with contact angles of 164.4° and 162.3° and sliding angles of less than 10°. Additionally, the coatings presented strong resistance to organic solvents and exhibited a significant photothermal effect [44].

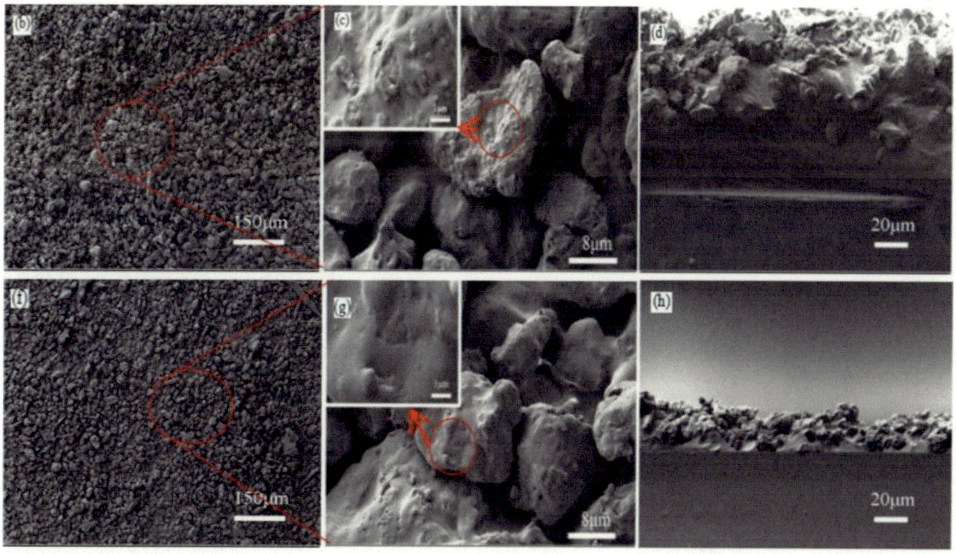

Figure 4.
SEM micrographs of surface morphology using L-CNC particles at high magnification and cross section. Reproduced with permission from [25] published by Materials, 2017.

4. Superhydrophobic coatings from proteins

4.1 Proteins

Proteins are biological macromolecules whose monomer units are α-amino acids sequence of the polypeptide chain, which play an important role in most of the biochemical functions of the cell. Their functionality depends on their structure and can be chemically, enzymatically, or physically modified due to the higher number of functional groups [45, 46]. Inspired by nature and enhancing interfacial adhesion, researchers are developing the use of proteins (i.e., Lysozyme) to provide nanoscale surface roughness [47, 48].

4.1.1 Zein

Zein is classified as a prolamin protein and is primarily derived from the endosperm of corn, accounting for 80% of its protein composition [49]. Its unique amino acid sequence gives it an amphiphilic nature. It has a high concentration of hydrophobic amino acids (50% hydrophobic residues), making it insoluble in water. However, it can be dissolved in the presence of organic solvents such as ethanol, high concentrations of urea and alkali (at pH 11), or anionic detergents (around 21–26% hydrophilic) [49, 50].

Recently, there has been a growing interest in zein as a renewable and biodegradable material to act as a coating agent for industries such as food packaging, biomedicine, and pharmaceuticals [51]. Researchers have been using the electrospinning technique to create zein fibers for the fabrication of superhydrophobic coatings [8]. These coatings have been found to exhibit contact angles above 150°. The concentration of zein determines the hydrophobicity of the surface. When the concentration is low, the surface becomes rough due to collapsed beads forming. However, when the concentration increases, the surface becomes smoother, forming fibers that reduce the contact angle [52]. Another eco-friendly method using zein is by producing zein nanoparticles that can enhance the antimicrobial activity of textiles [53]. To achieve this, ellagic acid was added as it is known for its effective antibacterial properties against *E. coli* and *S. aureus* (**Figure 5**) [54]. Despite the use of zein nanoparticles, the surface roughness was insufficient to attain superhydrophobicity, which requires contact angles of higher than 150° [53].

A dual-layered superhydrophobic coating can be fabricated through a cost-effective spray coating method by combining waxes' (beeswax and candelilla wax) low surface energy and roughness while zein/pectin nanoparticles and cellulose nanofibers act as a supporting layer. The coating possesses higher water repellence to different types of food liquids (tea, milk, honey, and coke) with water contact angles greater than 150° and sliding angles less than 20° [55].

4.1.2 Other proteins: Lysozyme and bovine serum albumin

Recent reports have proved that other types of proteins can be applied for superhydrophobic coatings. Lysozyme is a type of glycohydrolase that shows great potential. [48, 56]. It can hydrolyze the β-1,4-linkages found between N-acetylmuramic acid and N-acetyl-d-glucosamine residues in the peptidoglycan of bacterial cell walls. It is extracted from eggs and milk, and it is recognized as safe when in contact with food [57]. By utilizing cysteine as a reducing agent, the phase transition of lysozyme

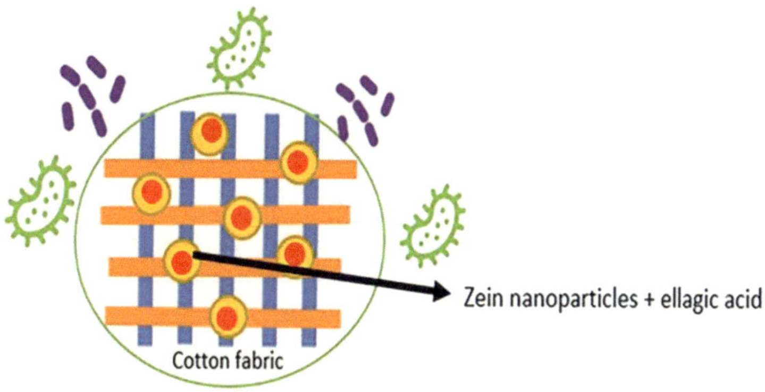

Figure 5.
Antimicrobial fabric with zein nanoparticles [53].

results in the production of micro/nano roughness that can effectively adhere to various surfaces. A durable and water-repellent coating (contact angles above 148° and sliding angles less than 10° depending on the food liquid) that resists adhesion to different types of food, such as yoghurt, milk, honey, and so forth, can be formed by using a combination of phase-transited lysozyme and carnauba wax. Additionally, to this, the coating showed high repellency to blood, and the blood droplet rolled easily without any residue [48].

In addition to this, it is possible to modify lysozyme (microparticles with 500 nm in diameter) to provide surface roughness by transitioning it into a microparticle-aggregated necklace network with internal amyloid stacking structures [8]. Lysozyme particles combined with a hydrophobic treatment turned different types of surface materials (cotton, ceramic, wood, PET, etc.) into superhydrophobic surfaces. It was demonstrated that lysozyme crystallization can be accelerated and facilitate protein-based superhydrophobic surfaces by using the facile protein spotting technique [47].

Bovine serum albumin is another type of protein composed of a single polypeptide that contains 583 amino acid residues [58]. Bovine serum albumin nanoparticles were synthesized by adding dipentaerythritol penta-acrylate to form covalently cross-linked nanoparticles in which bovine serum amine groups react with acrylate groups. The nanoparticles were deposited in cotton fibers achieving contact angles above 150° due to the rough structure and chemical modification, making them resistant to certain physical manipulations, extended UV radiation exposure, and chemical exposure [59].

5. Superhydrophobic coatings from lipids

5.1 Lipids

Lipids are hydrophobic and amphipathic molecules with high molecular weight. They are primarily soluble in organic solvents and not soluble in water. These structures act as energy storage within cells and offer thermal insulation [15]. Plants and algae primarily contain phosphoglycerides and glycosylglycerides as their main lipid components. When in contact with water, lipids exhibit a variety of behaviors such as forming monolayers, micelles, bilayers, membranes, or vesicles. This is due to water's strong cohesive self-attraction, which repels the hydrocarbon chains present in lipids [60].

In many organisms such as plants, lipids contribute to the surface covering including surface waxes, cutin, and suberin. Commonly, an example of this is the lotus effect exhibited by the leaves of the *Nembulo nucifera* flower [1]. The nanoscale rough structure combined with a natural hydrophobic wax helps with self-cleaning action. This action causes water droplets to roll into the surface, protecting the leaves from foreign organisms [61].

5.1.1 Waxes

Waxes consist of long-chain compounds that are typically nonpolar, meaning they are pure hydrocarbons or have a small hydrophilic component that does not easily interact with water, and they are highly hydrophobic in nature [60, 62, 63]. They can be categorized into natural and synthetic. Natural waxes are obtained from plants and animals [7]. Waxes such as carnauba wax, rice bran wax, soy wax, and candelilla wax are plant-derived, while beeswax is an example of a wax derived from animals [62, 63]. Additionally, petroleum waxes like paraffin and microcrystalline are also considered sustainable materials [6, 63, 64]. Waxes are soluble in organic solvents, solid at room temperature, and malleable [63, 65]. Their chemical composition makes them one of the most hydrophobic materials in nature. A significant amount of data indicates that waxes' chemical composition is complex and has not yet been fully established [62, 63]. The chemical composition of waxes varies depending on how the wax was extracted, the stage in which waxes were collected, and the environmental conditions [66]. They are mostly constituted by long-chain aliphatic compounds (10–60 carbon atoms) including fatty acids, hydrocarbons, alcohols, ketones, aldehydes, and esters (**Table 1**) [63].

Because waxes are economical, readily available, biodegradable, and can even be consumed, they could be a sustainable solution for industrial applications including the food, textiles, and paper industries [75]. Waxes are often utilized as materials with low surface energy. However, it is also recognized that waxes can cause the formation of rough structures because of the crystallization process, which creates crystals when

Wax	Chemical composition
Carnauba wax	Free fatty acids (3–3.5%), resins (4–6%), alcohols (2–3%), hydrocarbons (1–3%), and long-chain esters (84–85%) [67].
Candelilla wax	Hydrocarbons, free acids (7–9%), n-alkanes, esters, alcohols, sterols (20–29%), and resins (12–14%) [68].
Rice bran wax	Fatty acids (typically C20–C26) and fatty alcohols (C30–C36) [69, 70].
Beeswax	Esters (67%), free acids (12%), free alcohols, hydrocarbons (14%), and other non-identified components (6%) [71].
Soy wax	It is mainly composed of triacyl glyceride with a high portion of stearic acid. The wax came from soy oil, which consists of several fatty such as palmitic acid (10%), stearic acid (4%), oleic acid (18%), linoleic acid (55%), and linolenic acid (13%) [72, 73].
Paraffin wax	Alkanes (80–90%) with around 20–30 carbons [74].

Table 1.
Chemical composition of some biodegradable waxes based on references.

they are deposited in a substrate [76]. For this reason, waxes can create superhydrophobic surfaces by themselves combining both features [62, 77].

A superhydrophobic coating was made of candelilla wax and rice bran wax via a one-step spray solution. The waxes were dissolved in hot ethanol using temperature above their melting point. The coating presented micro/nanoroughness, which allowed it to repel several types of food liquids (milk, cola, orange juice, etc.) with WCA ~ 155° [78]. Another superhydrophobic coating was fabricated using soy wax dispersed in hot ethanol. The non-wettable coating had excellent repelling properties against different non-Newtonian food liquids with a higher apparent contact angle of 159° [73]. In addition to this, a general all-by-one superhydrophobic coating was made with food-grade waxes (beeswax and paraffin wax). The suspension was created using ethanol (95%) at 80°C and then ultrasonicated to homogenize the components. The superhydrophobic coating was applied to distinct types of food packaging materials obtaining water contact angles of ~160° and sliding angles of ~6.2° [76]. From the combination of candle soot particles and carnauba wax, a low-cost superhydrophobic surface can be fabricated. By depositing the concentrations of candle soot and carnauba wax into the glass substrate, the solvent evaporated completely leaving a rough structure with a contact angle of 172° [79].

A novel superhydrophobic coating with coffee lignin and beeswax (**Figure 6**). The wax wrapped the coffee lignin surface to increase hydrophobicity. Due to the low melting point of beeswax, the thermal stability was unsatisfactory at higher temperatures (120°C), causing a decrease in contact angle (from 165 to 89°C) [80]. Another approach using waxes as low surface energy material is by combining polylactic acid (PLA) and carnauba wax (CW). PLA/CW coating presented contact angles above 150° when it is in contact with different liquids. However, if the coating is exposed to a higher temperature (above the melting point of the wax), the coating without PLA starts to lose roughness while PLA/CW coating maintains superhydrophobic properties due to PLA rough structure [81]. Based on the results, it appears that using the spray coating technique is more effective than the dip coating technique. This is because it improves the roughness, which is necessary for achieving superhydrophobicity [81].

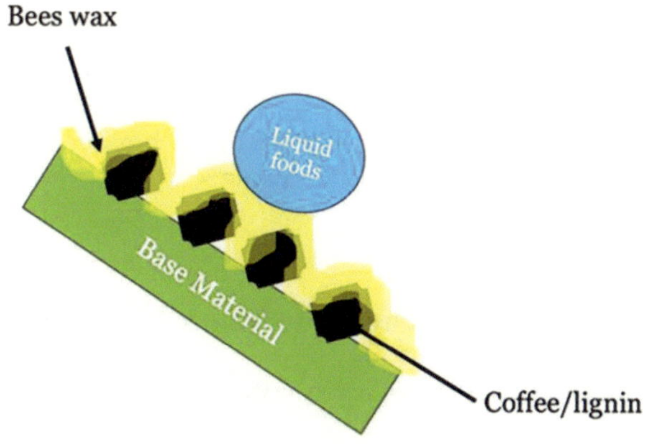

Figure 6.
Superhydrophobic coating prepared by coffee/lignin and beeswax. The coating has high repellence to different types of liquid foods [80].

Figure 7.
Stearic acid structural formulate [84].

5.2 Fatty acids

Fatty acids are part of the lipids and play an important role in metabolism. They are the main constituents of oils, fats, waxes, and lipid-based materials [82]. Fatty acids are carbon chains with a methyl group at the end of the molecule and a carboxyl group at the other end [83]. Most fatty acids are insoluble in water, which makes them hydrophobic in nature. Depending on the degree of saturation and length, the properties of the fatty acid vary [8]. Moreover, it is common for fatty acids to associate with one another, resulting in the creation of monolayers and micelles [83]. Stearic acid is a commonly used fatty acid for creating superhydrophobic surfaces and hydrophobic compounds [8]. Stearic acid ($CH_3(CH_2)_{16}COOH$) is a long-chain saturated fatty acid, and due to its non-toxicity and biocompatibility, it is used for cosmetics and pharmaceutical applications (**Figure 7**) [84, 85].

Inexpensive solutions for repellent surfaces to avoid water corrosion and another type of liquids for aluminum alloys are becoming attractive for industrial applications [8, 85, 86]. Applying the chemical etching technique, a superhydrophobic coating from stearic acid was made. The morphology of the coating shows micrometer and nanometer structures, which were able to trap enough air, preventing water penetration. The coating with a water contact angle of 156 ± 1° was more stable and resilient to corrosive liquids such as acids and bases compared to the coatings made by traditional techniques [86]. Another approach uses stearic acid and epoxidized soybean oil in cotton fabric etched by deep eutectic solvent. To improve the surface roughness of cotton fabric, the eutectic solvent was used while stearic acid and soybean oil coated the surface to provide low surface energy. A durable and flexible superhydrophobic cotton fabric (>150°) with self-cleaning properties and potential application for water separation was achieved [87]. An environmentally friendly and economical superhydrophobic coating was made from sepiolite nanoparticles functionalized with fatty acids (cinnamic acid and myristic acid) using the dip coating method. Cinnamic acid and myristic acid provide low surface free energy to sepiolite nanoparticles. The coatings showed a static contact angle of 160° with anti-biofouling and antimicrobial characteristics. A decrease of Gram-positive and -negative bacteria from 30% to 2–8% was observed [88].

6. Superhydrophobic coatings from other plant-based materials

Researchers are investigating alternative methods to create biodegradable superhydrophobic coatings. One possibility is using lycopodium spores, a type of plant-based material with a rough surface and naturally hydrophobic. By combining the spores' natural roughness with a material that has lower surface energy, it is possible to achieve superhydrophobicity [8].

6.1 Lycopodium spores

Lycopodium spores are plant-based materials that come from ground pine. These spores are intrinsically hydrophobic and present spherical shapes. They have a characteristic length scale of 20 μm, which provides surface roughness (**Figure 8**) [90]. Surfaces made from lycopodium spores have a higher water contact angle of 144° and stronger adhesion to surfaces, making them classified as "sticky" [91]. Lycopodium spores are commonly used as a carrier for oral drugs and vaccines [89]. These spores serve as natural protective packaging that offers high uniformity and large cavities to encase different materials, particularly for medical purposes [92].

Researchers are investigating new uses for lycopodium spores, particularly in creating superhydrophobic coatings. For example, a superhydrophobic surface was made by lycopodium spores to disperse in a Mater-Bi polymer matrix (a blend of vegetable-derived polyester with thermoplastic starch). Researchers achieved a water contact angle of approximately 150°. However, the coatings were sticky, preventing water droplets from moving freely across the surface, commonly referred to as the "rose petal effect" [93]. Recently, a superhydrophobic coating using lycopodium spores with a combination of two natural waxes (Carnauba wax and Beeswax) has been fabricated. The components were probe sonicated to correctly disperse the particles in water. As a result, they obtained contact angles above 155° and contact angles of hysteresis less than 7° [90].

7. The route of real-world application

Numerous types of plant-based materials have been reported in the literature for creating superhydrophobic coatings. Although these materials offer a range of useful functionalities, their commercial development has been limited. This could be attributed to different factors such as low physical resilience, durability, material compatibility, scaling up complications, and so forth [94]. Comparing the coatings

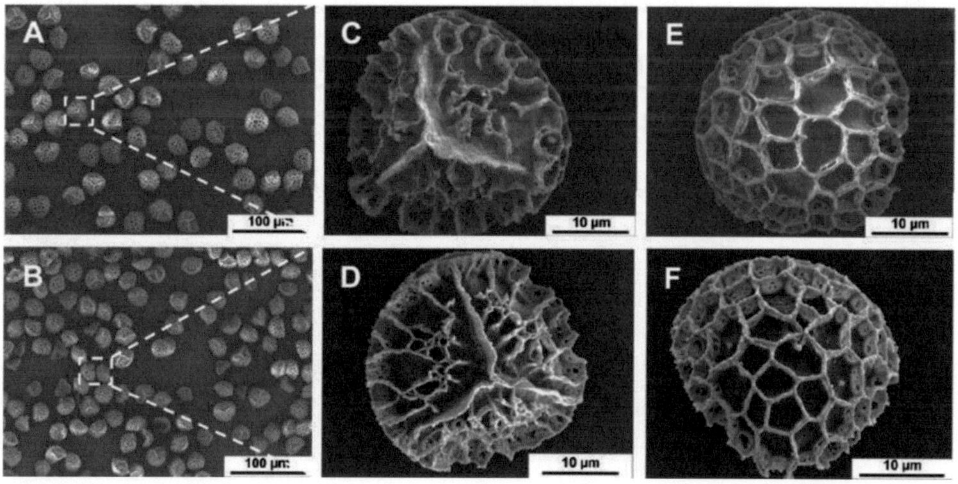

Figure 8.
SEM micrographs of lycopodium spores showing the honeycomb-like architecture. Reprinted (adapted) with permission from [89] Copyright 2023 American Chemical Society.

fabricated by synthetic materials and superhydrophobic coatings made from environmentally friendly materials, it can be inferred that both have been facing the same issues. Coatings derived from plant-based materials offer various benefits such as their ability to biodegrade, nontoxic nature, and reduced impact on the environment [8]. However, they are still needed chemical modification due to their fragility and hydrophilic behavior. These factors are important to consider for real-world applications to have a potential low-cost manufacturing process [4].

The global market of superhydrophobic coatings is incorporating more fluorine-free coatings. These commercial products are mainly used in nonmedical fields such as construction, automotive, aerospace, and marine [4]. Sectors such as food packaging aim to incorporate biodegradable coatings to minimize plastic pollution. This shift toward sustainability could also be implemented in the medical and textile sectors [3, 8]. Currently, to the best of our knowledge, there are no superhydrophobic coatings available in the market that are made solely from natural materials without any chemical modification.

8. Conclusion

In this chapter, we have outlined some affordable techniques and eco-friendly materials for fabricating superhydrophobic coatings. The use of biodegradable materials including biopolymers such as polysaccharides, proteins, and lipids as well as other plant-based materials including lycopodium showed tremendous potential for various applications. Based on the literature, it has been shown that many materials can function as rough agents by undergoing modifications that result in the formation of nanoparticles for the creation of rough surface textures. In addition, lipid compounds such as waxes can act as low surface energy materials; by combining both, it is possible to obtain superhydrophobicity.

Several low-cost and simple techniques (e.g., spray coating, dip coating, etc.) were used to fabricate such coatings. It is important to notice that superhydrophobic coatings from bio-based materials are still being developed to achieve better robustness and longevity, while avoiding using organic solvents and other chemical modifications, which involves toxic and hazardous compounds. Adding environmentally friendly materials would enhance the biodegradability and recyclability of the coatings.

Acknowledgements

BRD would like to acknowledge the Queen Mary University of London for funding.

Conflicts of interest

BRD would like to declare no conflict of interest.

Author details

Brenda Resendiz Diaz* and Colin R. Crick
School of Engineering and Materials Science, Queen Mary University of London, London, United Kingdom

*Address all correspondence to: b.resendiz-diaz@qmul.ac.uk

References

[1] Subhash Latthe S, Basavraj Gurav A, Shridhar Maruti C, et al. Recent progress in preparation of superhydrophobic surfaces: A review. Journal of Surface Engineered Materials and Advanced Technology. 2012;**02**:76-94

[2] Hooda A, Goyat MS, Pandey JK, et al. A review on fundamentals, constraints and fabrication techniques of superhydrophobic coatings. Progress in Organic Coatings. 2020;**142**. Epub ahead of print. DOI: 10.1016/j.porgcoat.2020.105557

[3] Ruzi M, Celik N, Onses MS. Superhydrophobic coatings for food packaging applications: A review. Food Packaging and Shelf Life. 2022;**32**. Epub ahead of print. DOI: 10.1016/j.fpsl.2022.100823

[4] Ghasemlou M, Daver F, Ivanova EP, et al. Bio-inspired sustainable and durable superhydrophobic materials: From nature to market. Journal of Materials Chemistry A. 2019;7:16643-16670

[5] Peng J, Wu L, Zhang H, et al. Research progress on eco-friendly superhydrophobic materials in environment, energy and biology. Chemical Communications. 2022;**58**:11201-11219

[6] Tharanathan RN. Biodegradable films and composite coatings: Past, present and future. Trends in Food Science and Technology. 2003;**14**:71-78

[7] Razavi SMR, Oh J, Sett S, et al. Superhydrophobic surfaces made from naturally derived hydrophobic materials. ACS Sustainable Chemistry & Engineering. 2017;**5**:11362-11370

[8] Bayer IS. Superhydrophobic coatings from ecofriendly materials and processes: A review. Advanced Materials Interfaces. 2020;7. Epub ahead of print. DOI: 10.1002/admi.202000095

[9] Jeevahan J, Chandrasekaran M, Britto Joseph G, et al. Superhydrophobic surfaces: A review on fundamentals, applications, and challenges. Journal of Coatings Technology and Research. 2018;**15**:231-250

[10] Cui C, Gao L, Dai L, et al. Hydrophobic biopolymer-based films: Strategies, properties, and food applications. Food Engineering Reviews. 2023;**15**:360-379

[11] Niaounakis M, editor. Definitions of terms and types of biopolymers. In: Biopolymers: Applications and Trends. William Andrew Publishing; 2015. pp. 1-90

[12] Shekar S, Ramachandra M. Green Composites: A Review. 2018. Available from: www.sciencedirect.comwww.materialstoday.com/proceedings2214-7853

[13] Sanivada UK, Mármol G, Brito FP, et al. PLA composites reinforced with flax and jute fibers—A review of recent trends, processing parameters and mechanical properties. Polymers (Basel). 2020;**12**:2373

[14] Weber CJ, Haugaard V, Festersen R, et al. Production and applications of biobased packaging materials for the food industry. Food Additives and Contaminants. 2002;**19**:172-177

[15] Jacob J, Gopi S. Chapter 3 - Isolation and physicochemical characterization of biopolymers. In: Thomas S, Gopi S, Amalraj A, editors. Biopolymers and their Industrial Applications. Cochin,

Kerala, India: Elsevier, R&D Centre, Aurea Biolabs (P) Ltd; 2021. pp. 45-79

[16] Coma V. Polysaccharide-based biomaterials with antimicrobial and antioxidant properties. Polimeros. 2013;**23**:287-297

[17] Alexandru G, Alina H. Biopolymers for Food Design. San Diego, CA, United States: Elsevier Inc.; 2018. [Epub ahead of print]. DOI: 10.1016/C2016-0-00686-1

[18] Lisitsyn A, Semenova A, Nasonova V, et al. Approaches in animal proteins and natural polysaccharides application for food packaging: Edible film production and quality estimation. Polymers. 2021;**13**. Epub ahead of print. DOI: 10.3390/polym13101592

[19] Wertz J-L, Bédué O, Mercier J-P. Cellulose Science and Technology. 1st ed. New York: EPFL Press English Imprint; 2010. p. 350. DOI: 10.1201/b16496. ISBN: 9780429131202

[20] O'Sullivan AC. Cellulose: The structure slowly unravels. Cellulose. 1997;**4**:173-207. DOI: 10.1023/A:1018431705579

[21] Wei DW, Wei H, Gauthier AC, et al. Superhydrophobic modification of cellulose and cotton textiles: Methodologies and applications. Journal of Bioresources and Bioproducts. 2020;**5**:1-15

[22] Teisala H, Tuominen M, Kuusipalo J. Superhydrophobic coatings on cellulose-based materials: Fabrication, properties, and applications. Advanced Materials Interfaces. 2014;**1**:1300026

[23] Wang Y, Zhao W, Han L, et al. Superhydrophobic surfaces from sustainable colloidal systems. Current Opinion in Colloid & Interface Science. 2022;**57**:101534

[24] Huang J, Lyu S, Fu F, et al. Green preparation of a cellulose nanocrystals/polyvinyl alcohol composite superhydrophobic coating. RSC Advances. 2017;**7**:20152-20159

[25] Huang J, Wang S, Lyu S. Facile preparation of a robust and durable superhydrophobic coating using biodegradable lignin-coated cellulose nanocrystal particles. Materials. 2017;**10**:1080

[26] Ye M, Wang S, Ji X, et al. Nanofibrillated cellulose-based superhydrophobic coating with antimicrobial performance. Advanced Composites and Hybrid Materials. 2023;**6**:30

[27] Geissler A, Chen L, Zhang K, et al. Superhydrophobic surfaces fabricated from nano- and microstructured cellulose stearoyl esters. Chemical Communications. 2013;**49**:4962

[28] Mourya VK, Inamdar NN. Chitosan-modifications and applications: Opportunities galore. Reactive and Functional Polymers. 2008;**68**:1013-1051

[29] Kaushal N, Singh AK. Advancement in utilization of bio-based materials including cellulose, lignin, chitosan for bio-inspired surface coatings with special wetting behavior: A review on fabrication and applications. International Journal of Biological Macromolecules. 2023;**246**:125709

[30] Roy S, Goh K-L, Verma C, et al. A facile method for processing durable and sustainable superhydrophobic chitosan-based coatings derived from waste crab shell. ACS Sustainable Chemistry & Engineering. 2022;**10**:4694-4704

[31] Tagliaro I, Seccia S, Pellegrini B, et al. Chitosan-based coatings with tunable transparency and

superhydrophobicity: A solvent-free and fluorine-free approach by stearoyl derivatization. Carbohydrate Polymers. 2023;**302**:120424

[32] Appelqvist IAM, Debet MRM. Starch-biopolymer interactions-A review. Food Reviews International. 1997;**13**:163-224

[33] Apriyanto A, Compart J, Fettke J. A review of starch, a unique biopolymer–Structure, metabolism and in planta modifications. Plant Science. 2022;**318**:111223

[34] Wang F, Chang R, Ma R, et al. Eco-friendly and pH-responsive nano-starch-based superhydrophobic coatings for liquid-food residue reduction and freshness monitoring. ACS Sustainable Chemistry & Engineering. 2021;**9**:10142-10153

[35] Wang F, Chang R, Ma R, et al. Eco-friendly and superhydrophobic nano-starch based coatings for self-cleaning application and oil-water separation. Carbohydrate Polymers. 2021;**271**:118410

[36] Wang F, Ma R, Tian Y. Facile fabrication of thermostable and colorimetric starch-based waterproof coating with edible organic materials. Food Chemistry. 2022;**382**:132269

[37] Niu S, He W, Chang Q, et al. Beeswax-modified starch–cellulose composites with superhydrophobic and self-cleaning capability. Starch-Stärke. 2023;**75**. Epub ahead of print. DOI: 10.1002/star.202200285

[38] Yang Y, Deng Y, Tong Z, et al. Renewable lignin-based xerogels with self-cleaning properties and superhydrophobicity. In: ACS Sustainable Chemistry and Engineering. American Chemical Society. Guangzhou, People's Republic of China: School of Chemistry and Chemical Engineering, South China University of Technology, Research Institute of Materials Science; 2014. pp. 1729-1733

[39] Vanholme R, Morreel K, Ralph J, et al. Lignin engineering. Current Opinion in Plant Biology. 2008;**11**:278-285

[40] Vanholme R, Demedts B, Morreel K, et al. Lignin biosynthesis and structure. Plant Physiology. 2010;**153**:895-905

[41] Heber A, Alexandre N, Marcos M. Lignin structure and wood properties. Wood and Fiber Science. 1999;**31**:426-433

[42] Shorey R, Gupta A, Mekonnen TH. Hydrophobic modification of lignin for rubber composites. Industrial Crops and Products. 2021;**174**. Epub ahead of print. DOI: 10.1016/j.indcrop.2021.114189

[43] Ren C, Li M, Huang W, et al. Superhydrophobic coating with excellent robustness and UV resistance fabricated using hydrothermal treated lignin nanoparticles by one-step spray. Journal of Materials Science. 2022;**57**:18356-18369

[44] Ma B, Xiong F, Wang H, et al. Tailorable and scalable production of eco-friendly lignin micro-nanospheres and their application in functional superhydrophobic coating. Chemical Engineering Journal. 2023;**457**:141309

[45] Gómez-Estaca J, Gavara R, Catalá R, et al. The potential of proteins for producing food packaging materials: A review. Packaging Technology and Science. 2016;**29**:203-224

[46] Kent SBH. Total chemical synthesis of proteins. Chemical Society Reviews. 2009;**38**:338-351

[47] Gao A, Wu Q, Wang D, et al. A superhydrophobic surface templated

by protein self-assembly and emerging application toward protein crystallization. Advanced Materials. 2016;**28**:579-587

[48] Li J, Tian J, Gao Y, et al. All-natural superhydrophobic coating for packaging and blood-repelling materials. Chemical Engineering Journal. 2021;**410**:128347

[49] UNIVERSITÀ DEGLI STUDI DI TRIESTE UNIVERSITÀ CA' FOSCARI DI VENEZIA XXXIV CICLO DEL DOTTORATO DI RICERCA IN CHIMICA DOTTORATO IN COLLABORAZIONE CON ISTITUTO ITALIANO DI TECNOLOGIA ZEIN-BASED HYDROPHOBIC COATING: FROM PROMISING PROPERTIES TO FIELD APPLICATION Settore scientifico-disciplinare: CHIM/12-CHIMICA DELL'AMBIENTE E DEI BENI CULTURALI.

[50] Shukla R, Cheryan M. Zein: The industrial protein from corn. Industrial Crops and Products. 2001;**13**:171-192

[51] Corradini E, Curti P, Meniqueti A, et al. Recent advances in food-packing, pharmaceutical and biomedical applications of zein and zein-based materials. International Journal of Molecular Sciences. 2014;**15**:22438-22470

[52] Dong F, Zhang M, Tang W-W, et al. Formation and mechanism of superhydrophobic/hydrophobic surfaces made from amphiphiles through droplet-mediated evaporation-induced self-assembly. The Journal of Physical Chemistry B. 2015;**119**:5321-5327

[53] Gonçalves J, Torres N, Silva S, et al. Zein impart hydrophobic and antimicrobial properties to cotton textiles. Reactive and Functional Polymers. 2020;**154**:104664

[54] Ghudhaib KK, Hanna ER, Jawad AH. Effect of ellagic acid on some types of pathogenic bacteria. Journal of Al-Nahrain University-Science. 2010;**13**:79-85

[55] Wang T, Zhao Y. Fabrication of thermally and mechanically stable superhydrophobic coatings for cellulose-based substrates with natural and edible ingredients for food applications. Food Hydrocolloids. 2021;**120**:106877

[56] Liu S, Zheng J, Hao L, et al. Dual-functional, superhydrophobic coatings with bacterial anticontact and antimicrobial characteristics. ACS Applied Materials & Interfaces. 2020;**12**:21311-21321

[57] Syngai GG, Ahmed G. Chapter 11 - Lysozyme: A natural antimicrobial enzyme of interest in food applications. In: Kuddus M, editor. Enzymes in Food Biotechnology. Academic Press; 2019. pp. 169-179

[58] Jahanban-Esfahlan A, Ostadrahimi A, Jahanban-Esfahlan R, et al. Recent developments in the detection of bovine serum albumin. International Journal of Biological Macromolecules. 2019;**138**:602-617

[59] Shome A, Rather AM, Manna U. Chemically reactive protein nanoparticles for synthesis of a durable and deformable superhydrophobic material. Nanoscale Advances. 2019;**1**:1746-1753

[60] Callegarin F, Quezada GJA, Debeaufort F, et al. Lipids and biopackaging. Journal of the American Oil Chemists' Society. 1997;**74**:1183-1192. DOI: 10.1007/s11746-997-0044-x

[61] Crick CR. The chemistry and CVD of hydrophobic surfaces. [Doctoral thesis]. England: University College London (UCL); 2011

[62] Ribas GM. Luleå University of Technology, Department of Engineering

Sciences and Mathematics. [Independent thesis] Advanced level (degree of Master (Two Years)), 20 credits / 30 HE credits. [Student thesis]. National Category Engineering and Technology Identifiers; 2019

[63] Saji VS. Wax-based artificial superhydrophobic surfaces and coatings. Colloids and Surfaces A: Physicochemical and Engineering Aspects. 2020;**602**. Epub ahead of print. DOI: 10.1016/j.colsurfa.2020.125132

[64] El Assimi T, Lakbita O, El Meziane A, et al. Sustainable coating material based on chitosan-clay composite and paraffin wax for slow-release DAP fertilizer. International Journal of Biological Macromolecules. 2020;**161**:492-502

[65] Krendlinger EJ, Wolfmeier UH. Natural and Synthetic Waxes: Origin, Production, Technology, and Applications. Weinheim, Germany: WILEY-VCH GmbH; 2023

[66] Wang L, Ando S, Ishida Y, et al. Quantitative and Discriminative Analysis of Carnauba Waxes by Reactive Pyrolysis-GC in the Presence of Organic Alkali Using a Vertical Microfurnace Pyrolyzer. 2001. Available from: www.elsevier.com/locate/jaap

[67] de Freitas CAS, de Sousa PHM, Soares DJ, et al. Carnauba wax uses in food–A review. Food Chemistry. 2019;**291**:38-48

[68] Aranda-Ledesma NE, Bautista-Hernández I, Rojas R, et al. Candelilla wax: Prospective suitable applications within the food field. Lebensmittel-Wissenschaft & Technologie. 2022;**159**:113170

[69] Dassanayake LSK, Kodali DR, Ueno S, et al. Physical properties of rice bran wax in bulk and organogels. Journal of the American Oil Chemists' Society. 2009;**86**:1163

[70] Wijarnprecha K, Aryusuk K, Santiwattana P, et al. Structure and rheology of oleogels made from rice bran wax and rice bran oil. Food Research International. 2018;**112**:199-208

[71] Svečnjak L, Chesson LA, Gallina A, et al. Standard methods for Apis mellifera beeswax research. Journal of Apicultural Research. 2019;**58**:1-108

[72] Shaharuddin SIS, Mansor MA, Rashid MA, et al. The effects of beeswax additions on the structural, thermal, mechanical and mass loss properties of soy wax blends. Journal of Mechanical Engineering. 2023;**20**:61-75

[73] Shen T, Fan S, Li Y, et al. Preparation of edible non-wettable coating with soybean wax for repelling liquid foods with little residue. Materials. 2020;**13**. Epub ahead of print. DOI: 10.3390/ma13153308

[74] Palou A, Cruz J, Blanco M, et al. Characterization of the composition of paraffin waxes on industrial applications. Energy and Fuels. 2014;**28**:956-963

[75] Torun I, Ruzi M, Er F, et al. Superhydrophobic coatings made from biocompatible polydimethylsiloxane and natural wax. Progress in Organic Coating. 2019;**136**. Epub ahead of print. DOI: 10.1016/j.porgcoat.2019.105279

[76] Zhao X, Hu T, Zhang J. Superhydrophobic coatings with high repellency to daily consumed liquid foods based on food grade waxes. Journal of Colloid and Interface Science. 2018;**515**:255-263

[77] Wang W, Lockwood K, Boyd LM, et al. Superhydrophobic coatings with

edible materials. ACS Applied Materials & Interfaces. 2016;**8**:18664-18668

[78] Liu BY, Xue CH, An QF, et al. Fabrication of superhydrophobic coatings with edible materials for super-repelling non-Newtonian liquid foods. Chemical Engineering Journal. 2019;**371**:833-841

[79] Celik N, Kiremitler NB, Ruzi M, et al. Waxing the soot: Practical fabrication of all-organic superhydrophobic coatings from candle soot and carnauba wax. Progress in Organic Coating. 2021;**153**:106169

[80] Zhang Y, Bi J, Wang S, et al. Functional food packaging for reducing residual liquid food: Thermo-resistant edible super-hydrophobic coating from coffee and beeswax. Journal of Colloid and Interface Science. 2019;**533**:742-749

[81] Wang X, Chen K, Liu Y, et al. Preparation and application of biodegradable and superhydrophobic polylactic acid/carnauba wax coating. Progress in Organic Coating. 2023;**177**:107434

[82] Kenar JA, Moser BR, List GR. Naturally occurring fatty acids. In: Ahmad MU, editor. Fatty Acids. Peoria, IL, United States: National Center for Agricultural Utilization Research; Washington, IL, United States: Elsevier, AOCS Press; 2017. pp. 23-82. DOI: 10.1016/B978-0-12-809521-8.00002-7

[83] Rustan AC, Drevon CA. Fatty acids: Structures and properties. In: Encyclopedia of Life Sciences. Oslo, Norway: John Wiley & Sons, Ltd., University of Oslo; 2005. DOI: 10.1038/npg.els.0003894. Available from: www.els.net

[84] Mauri O. 3: Final report on the safety assessment of oleic acid, lauric acid, palmitic acid, myristic acid, and stearic acid. Journal of the American College of Toxicology. 1987;**6**:321-401

[85] Ng WF, Wong MH, Cheng FT. Stearic acid coating on magnesium for enhancing corrosion resistance in Hanks' solution. Surface and Coating Technology. 2010;**204**:1823-1830

[86] Wang Q, Zhang B, Qu M, et al. Fabrication of superhydrophobic surfaces on engineering material surfaces with stearic acid. Applied Surface Science. 2008;**254**:2009-2012

[87] Xu QB, Wang XY, Wang P, et al. Sustainable and superhydrophobic coating from epoxidized soybean oil and stearic acid on cotton fabric etched by deep eutectic solvent. Materials Today Chemistry. 2022;**26**:101211

[88] Razavi SMR, Oh J, Haasch RT, et al. Environment-friendly antibiofouling superhydrophobic coatings. ACS Sustainable Chemistry & Engineering. 2019;**7**:14509-14520

[89] Uddin MJ, Abidi N, Warzywoda J, et al. Investigation of the fate of proteins and hydrophilicity/hydrophobicity of lycopodium clavatum spores after organic solvent-base-acid treatment. ACS Applied Materials & Interfaces. 2019;**11**:20628-20641

[90] Morrissette JM, Carroll PJ, Bayer IS, et al. A methodology to produce eco-friendly superhydrophobic coatings produced from all-water-processed plant-based filler materials. Green Chemistry. 2018;**20**:5169-5178

[91] Bormashenko E, Stein T, Pogreb R, et al. 'Petal effect' on surfaces based on lycopodium: High-stick surfaces demonstrating high apparent contact angles. Journal of Physical Chemistry C. 2009;**113**:5568-5572

[92] Mundargi RC, Potroz MG, Park S, et al. Lycopodium spores: A naturally manufactured, superrobust biomaterial for drug delivery. Advanced Functional Materials. 2016;**26**:487-497

[93] Milionis A, Ruffilli R, Bayer IS. Superhydrophobic nanocomposites from biodegradable thermoplastic starch composites (Mater-Bi®), hydrophobic nano-silica and lycopodium spores. RSC Advances. 2014;**4**:34395-34404

[94] Mehanna YA, Sadler E, Upton RL, et al. The challenges, achievements and applications of submersible superhydrophobic materials. Chemical Society Reviews. 2021;**50**:6569-6612

Chapter 2

Perspective Chapter: Challenges in the Durability of Superhydrophobic Coatings Pertinent to Unmanned Aerial Vehicle (UAV) Icing Mitigation

Zichen Zhang, Junfei Ou and Wen Li

Abstract

Icing is a severe hazard to the flight safety of unmanned aerial vehicles (UAV) in cold climates. Great efforts have been put into developing a reliable and practical anti-/de-icing system for icing mitigation. Compared to the conventional thermal-based deicing systems, applying superhydrophobic coatings with excellent ice-repellent performance is a promising anti-icing technique with low weight and low energy input. However, the application of superhydrophobic coatings on UAV is hindered by the substandard durability. This chapter comprehensively reviews advances in durability tests of superhydrophobic coatings pertinent to UAV anti-/de-icing. This chapter specifically introduces the mechanism of degrading superhydrophobic coatings by droplet erosion, which is rarely discussed in previous studies but can cause severe damage to the surface topology by applying "water hammer pressure." Consequently, recommendations are provided to facilitate a more systematic approach to conducting durability tests for superhydrophobic coatings.

Keywords: durability, superhydrophobic coating, anti-icing, droplet erosion, aircraft icing

1. Introduction

Aircraft icing plays a critical factor in endangering flight safety, since it can distort airflow over the wing, degrade the controllability, cause the stall at a lower angle of attack and lead to roll and pitch upsets [1]. Federal Aviation Administration (FAA), European Union Aviation Safety Agency and Civil Aviation Administration of China have issued strict icing airworthiness regulations because in-flight icing has resulted in several severe aviation accidents [2]. For example, American Eagle Flight 4184 crashed because of icing, resulting in 68 fatalities; and in 2004, China Eastern Airlines Flight MU5210 crashed due to wing icing (or frost), resulting in the loss of 55 lives. Data disclosed by FAA show that icing incidents led to 46 flight safety accidents between 2008 and 2016 [3]. The development of unmanned aerial vehicles (UAV) has indeed seen rapid growth in recent years, which is attributed to the increasing demand in the civil and military applications. However, during military operations, 25% of UAV flights encountered icing, which had a detrimental impact on mission success [4]. Common strategies for UAV icing avoidance involve either grounding the UAV or modifying path planning, which significantly reduces the operational capabilities of the UAV in cold climates [5].

Advanced anti-/de-icing systems are essential protections for aircraft in adverse weather conditions. Most airliners rely on the thermal deicing system, which involves heating the icing components during flight to remove the accreted ice. The thermal de-icing method uses either hot air from the engine or electrical heating parts to melt the accreted ice [6]. High-voltage plasma deicing is also a promising thermal deicing method, which can achieve the flow and icing control with the same equipment [7]. However, these anti-/de-icing techniques may be too complex, heavy and power-intensive to be practical for UAV, and therefore, may not be suitable for small-scale, lightweight UAV due to payload limitations and limited available power [8]. As a result, a hybrid deicing system combining a thermal deicing system on the wing/propeller's leading edge and a superhydrophobic coating covering the wing surface was developed as a low-power hybrid deicing strategy [9]. This hybrid deicing method has been used on unmanned aviation vehicles and wind turbines. As a part of the hybrid deicing system, superhydrophobic coating is a promising anti-icing technology with lightweight, low energy consumption and excellent water/ice repellency [10]. Because of the air pockets underneath the water droplet or ice, some superhydrophobic coatings can significantly reduce ice adhesion by reducing surface energy and delaying icing time. Microcraks between the solid surface and the water or ice may be introduced by the air pockets. The size of the microcracks is the key factor that governs ice adhesion. Consequently, some superhydrophobic surfaces significantly reduce ice adhesion if they provide sufficiently large microcraks at the interface.

Superhydrophobic coatings can be fabricated in many ways, but the primary approach is either patterning roughness on the solid surface, followed by attaching a thin layer of a low surface energy material or depositing materials with nanostructures and low surface energy on the surface. Fluoropolymers or siloxanes are classical materials with low surface energy employed to achieve superhydrophobicity [11]. Common techniques to fabricate superhydrophobic coatings include lithography, sol-gel method, electrochemical deposition, chemical vapor deposition, spraying, imprinting, hydrothermal method and phase separation [12]. Yang et al. [13] reported a facile phase separation route to produce anti-corrosion superhydrophobic polyvinylchloride coatings on a magnesium alloy (see **Figure 1(a)**). Ding et al. [14] described the construction of the superhydrophobic surface by electrospinning the poly(vinyl alcohol) and poly(vinyl alcohol)/zinc acetate solutions followed by the calcination process, which yields fibrous zinc oxide superhydrophobic films (see **Figure 1(b)**). The wetting properties of the fibrous films were further altered by applying a coating of a low surface energy fluoroalkylsilane using hexane. **Figure 1(c)** shows the schematic diagram of the reproduction route to create a robust superhydrophobic coating using the nanoimprint technique for rose petals [15]. Singh et al. [17] reported the synthesis of superhydrophobic coating on cotton fabrics using a sol-gel route (see **Figure 1(d)**). In addition to using a single method, Wang et al. [18] obtained a highly durable superhydrophobic coating by combining lithography and spraying techniques. Several bulk materials can also provide superhydrophobic surfaces. Zhu et al. [16] developed a superhydrophobic bulk material by embossing carbon nanotubes on

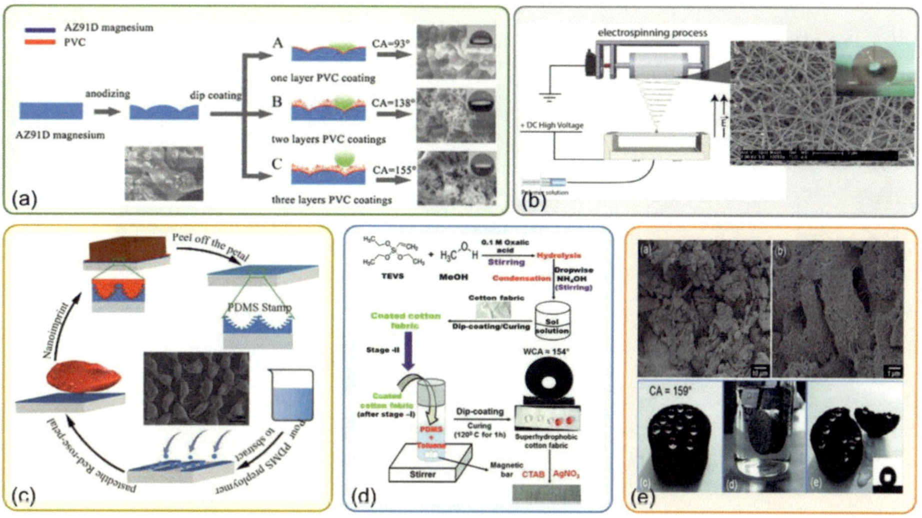

Figure 1.
Techniques to fabricate superhydrophobic coatings: (a) phase separation method, (b) electro-spinning method, (c) nano-imprinting method, (d) hydrothermal method and (e) a superhydrophobic bulk material [13–16].

polytetrafluoroethylene mold to create the necessary hierarchical texture required for the superhydrophobicity (see **Figure 1(e)**). Some hydrophilic materials can also be used to fabricate superhydrophobic coatings. Li et al. [19] employed plasma etching and plasma deposition of hydrophilic diamond-like carbon to a paper and successfully got a superhydrophobic surface. The mechanism is that the hierarchical nanostructure of the carbon provides an appropriate roughness to reduce the surface energy.

While some superhydrophobic coatings that can provide large microcraks have excellent icephobic performance, none can be used in anti-/de-icing systems of aircraft due to the limited durability. As early as 1969, the National Aeronautics and Space Administration (NASA) collaborated with the FAA to test 100 icephobic coatings in the icing wind tunnel at Glenn Research Center, but none could be used for aircraft applications [20]. NASA then established Revolutionary Icing Materials Evaluation Laboratory (RIMELab, previously referred to as Fundamental Adhesion and Shedding Test laboratory) at the Glenn Research Center, which is equipped with advanced ice adhesion measurement and research equipment. RIMELab is dedicated to developing low-ice-adhesion coatings, testing coating durability and quantifying the ice adhesion strength.

In this way, enhancing the durability of superhydrophobic coatings is crucial for application in aircraft. NASA also highlighted that the development of superhydro-/ice-phobic coatings is mainly hindered by the coating durability subject to the operational environment [21]. As an aeronautic superhydrophobic coating, it should perform well in environments with high humidity, high velocity, low temperature, thermal shocks, mechanical wear, corrosion, intense radiation and high-speed water droplet impacts [22]. As a result, the primary types of durability performance of superhydrophobic coatings are droplet/particle erosion resistance, mechanical abrasion resistance, icing/deicing cycles durability, UV resistance and corrosion resistance [23]. Droplet erosion tests (see **Figure 2(a)**) simulate the impact of micro-droplets or rain droplets, which could gradually degrade the wettability of superhydrophobic coatings due to a Cassie to Wenzel transition. Droplet erosion could be as destructive

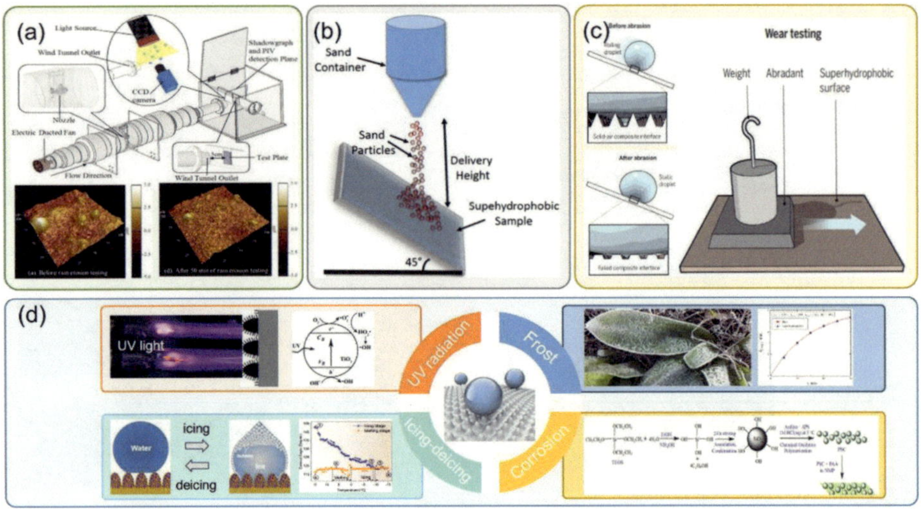

Figure 2.
Classification of the durability of superhydrophobic coatings: (a) droplet erosion test, (b) sand erosion test, (c) mechanical abrasion test, (d) other durability tests [23–29].

as solid particle erosion because of the "water hammer effect" [26, 30]. The water hammer effect in droplet impact refers to the phenomenon where a sudden increase in pressure occurs upon the impact of a droplet on a surface. This effect is caused by the rapid deceleration of the droplet upon impact, leading to a sharp increase in pressure that propagates through the fluid. It can result in the generation of shock waves and high local pressures.

Particle erosion is a common test performed to evaluate the mechanical durability of superhydrophobic coatings. **Figure 2(b)** shows a typical experimental setup where particles impinge on the coating with an inclination angle of 45°. These collisions could cause substantial changes to the surface topography, leading to severe destructions of superhydrophobic performance. Mechanical wear is one of the most straightforward and common types of degradation for superhydrophobic coatings; therefore, it is important for characterizing durability [31]. Experiments of mechanical wear have a common concept of rubbing or sliding a solid material into the coating (see **Figure 2(c)**). Mechanical wear induces material removal that eventually leads to surface topology modification and superhydrophobicity loss. Ultraviolet (UV) light from sunlight could lead to photo-oxidation of superhydrophobic coatings and subsequently the increase of surface energy eventually leads to wettability degradation [32, 33]. Similarly, chemical corrosion, icing-deicing cycle and frost could greatly affect the surface topography and surface energy of superhydrophobic coatings [34].

Plenty of studies have introduced methods of testing or enhancing the durability of superhydrophobic coatings in recent years. However, there is a lack of comprehensive review covering these methods. As a result, this chapter comprehensively summarizes cutting-edge techniques to quantify the surface durability and fabricate durable superhydrophobic coatings. In the following context, the thermodynamics of the surface will be introduced at first to provide the fundamental theory of the wetting phenomenon. Then, methods to quantify the mechanical durability (e.g., sand erosion, droplet erosion and abrasion durability) of superhydrophobic coatings are demonstrated. State-of-art durable coatings that can effectively resist mechanical damage are also introduced. After a comprehensive review of the environmental factors affecting the durability of superhydrophobic coatings, promising approaches to improve the coating durability will be discussed aiming at inspiring great ideas for developing durable, user-friendly and cost-effective superhydrophobic coatings.

2. Thermodynamics of superhydrophobic coatings

Static, receding and advancing contact angles are significant parameters to quantify the wettability of a coating. The contact angle is the angle between the surface tangent on the liquid-vapor interface and the tangent on the solid-liquid interface at their intersection. The static contact angle is the result of the quasi-equilibrium between the liquid, the solid and the vapor phases, so its value on a smooth and homogeneous surface can be given by the Young equation [35]:

$$\gamma_{sv} - \gamma_{sl} = \gamma_{\lg} \cos\theta_Y \tag{1}$$

where γ_{sv}, γ_{sl} and γ_{lv} are solid-vapor, solid-liquid and liquid-vapor interfacial energies. θ_Y is the Young contact angle. Superhydrophobic coatings have a static contact angle above 150°. The static contact angle is independent of the inclination of the surface but changes with temperature and surface tension [36]. Eq. (1) can be related to the work of adhesion by the Young-Dupre equation:

$$\gamma_{\lg}\left(1+\cos\left(\theta_{rec}\right)\right) = W \tag{2}$$

where W is the work of adhesion, which is the amount of work required to create a new interface [37]. Eq. (2) demonstrates that droplets on superhydrophobic coatings (with a high static contact angle) overcome small work of adhesion in the wetting/dewetting process, so droplets on those surfaces can move easily [38].

While contact angle can quantify the equilibrium state of droplets, it is not an ideal parameter to predict the friction of droplets. For example, surfaces with "rose-petal-effect" have both high contact angle, high water adhesion and low receding angle, meaning droplets can hardly move on it. As a result, contact angle hysteresis is widely used to quantify the friction of droplets [39, 40]. Contact angle hysteresis (CAH) is defined as the difference between the advancing (θ_a, contact angle of an expanding droplet) and receding contact angles (θ_r, contact angle of a shrinking droplet). A widely used experimental setup of measuring the static contact angle, receding and advancing angles is shown in **Figure 3** [41–43]. The setup consists of a light source, a syringe and a camera that records the shape of droplets with a sharp edge. A syringe pump usually manipulates the syringe to control the water injecting/sucking rate precisely.

Most superhydrophobic coatings are rough and air gets trapped in their hierarchical structures (named as Cassie-Baxter state), thereby, Eq. (1) needs to be revised as:

$$\cos\theta_{CB} = f cos\theta + f - 1 \tag{3}$$

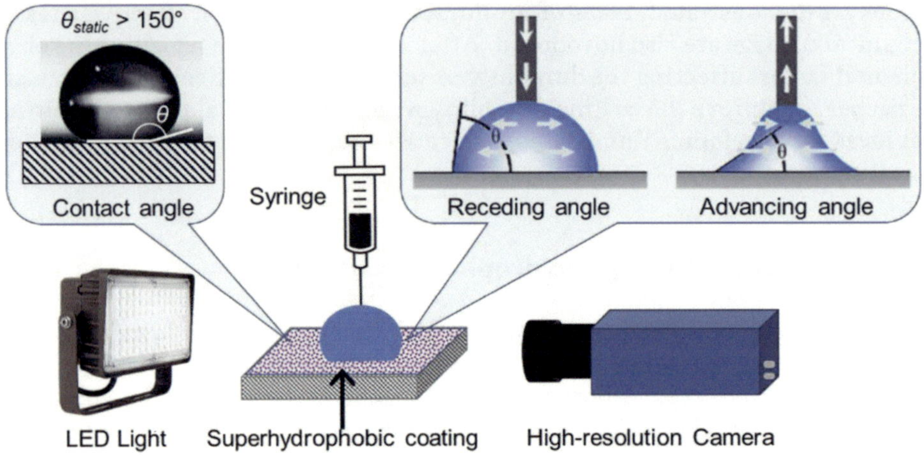

Figure 3.
The experimental setup measures the static, receding and advancing contact angles.

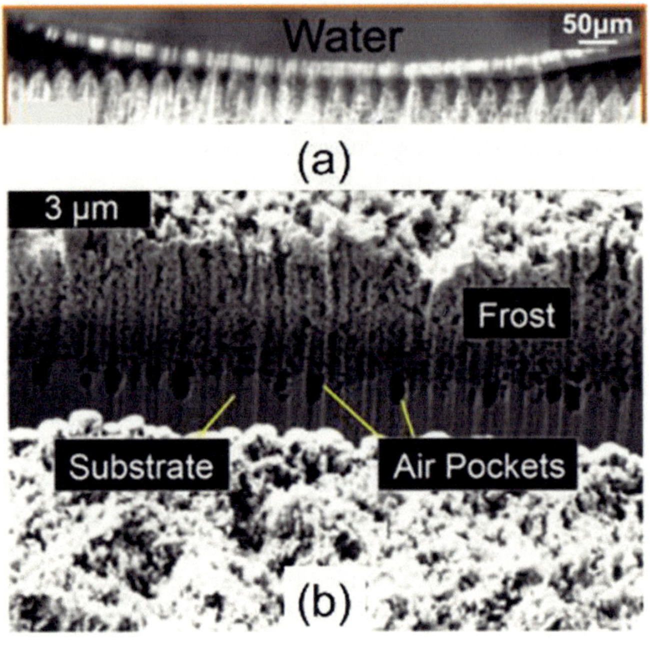

Figure 4.
Experimental evidence of Cassie-Baxter state: (a) a water droplet and (b) a frost layer on a superhydrophobic coating with visible air pockets [29, 44].

where f is the solid area fraction.

Recently, the Cassie-Baxter state has been observed in practice by cryo-SEM (see **Figure 4**) [29, 44]. **Figure 4(a)** clearly shows that sessile water droplets cannot penetrate the hierarchical structures on the superhydrophobic coating.

3. Mechanical durability of superhydrophobic coatings

3.1 Droplet erosion

Droplet erosion resistance is one of the most important durability for superhydrophobic coatings pertinent to aircraft/wind turbine icing. Droplet erosion indicates continuous impacts of rain droplets or high-speed airborne droplets [45]. For a representative case of a working offshore wind turbine coated with SHS as an anti-icing coating, rain droplets or sea spray droplets would impinge onto the exposed surface at a high relative velocity of around 80 m/s [46]. Some impinged water droplets may penetrate the micro-scale surface textures of superhydrophobic coatings, which refers to the wetting transition from the non-wetted Cassie-Baxter state to the Wenzel state, thereby, eliminating the hydrophobicity and icephobicity of the SHS. Furthermore, due to the continuous impacts of high-speed droplets, the surface topology of superhydrophobic coatings can be considerably altered and even some materials are worn away from the substrate [26]. It is well known that the superhydrophobicity of a surface is usually achieved by adding hydrophobicity agents onto micro-/nano-textured surfaces [47]. Therefore, superhydrophobic coatings with substantial surface topology change induced by rain erosion may lose the superhydrophobicity.

It should be noted that while extensive studies have been undertaken to investigate droplet erosion effects on different metal surfaces or special erosion-resistant coatings with extremely high-speed drop impacts (i.e., >200 m/s) [48], few studies have discussed the rain erosion mechanisms and effects on the wettability degradation (e.g., increasing contact angle hysteresis) on superhydrophobic coatings. Most of the previous studies about droplet erosion only focused on material or mass loss with extremely high-speed droplet impacts (i.e., >200 m/s). However, droplet-erosion-induced wettability degradation on superhydrophobic coatings occurs with a relatively lower droplet impingement speed (i.e., <100 m/s) without apparent material loss [26]. The water hammer effect in the early stage of the impact was found to be responsible for most of the damage resulting from droplet erosion [49]. In our previous study, we reported the wettability and icephobicity degradation as well as the micro-scale surface topology change upon superhydrophobic coatings after experiencing droplet erosion tests (i.e., impact velocity around 100 m/s), and illustrating the underlying physics of droplet erosion damage by analyzing the fundamentals of the droplet impact on a solid surface [26].

The fundamentals of droplet impact over a solid surface need to be addressed for exploring the mechanisms of rain erosion damage on superhydrophobic coatings. As described by Field [50], the whole impact process of a droplet can be divided into a compressible regime and an incompressible regime based on the progress of shock wave in the droplet. In the compressible regime, **Figure 5(a)** shows the initial compressible stage where the periphery of the contact area moves outward a short time after impact with a velocity higher than the speed of sound in the liquid and solid. It indicates that a shock wave envelope, after which droplet behaves compressible, is generated immediately giving tremendous pressure to the solid surface. This pressure is usually called "water hammer pressure" whose magnitude can be estimated using Eq. (4) where c_L and ρ_L are speed of sound and density of the liquid droplet [50]. However, liquid moving faster than the speed of sound near the periphery of the contact area even has a higher pressure reaching up to three times to the water hammer pressure based on the description of Lesser's geometrical acoustics model [51].

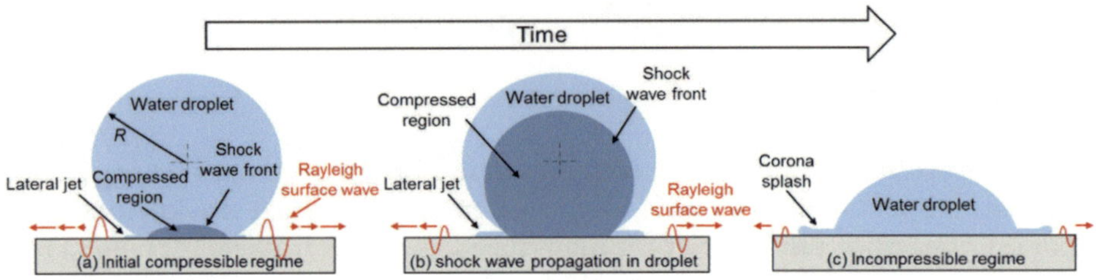

Figure 5.
Different stages of droplet impact on a solid surface. (a). Shock wave generation at the initial compressible regime. (b). Shock wave propagation in the droplet. (c). Incompressible regime.

The lateral jet shown in **Figure 5(a)** is liquids moving faster than the speed of sound in a droplet. As shown in **Figure 5(b)**, the shock wave can propagate in the droplet until it reaches the free surface where the release wave starts moving into the droplet and terminates the compressible regime after a short time. After the compressible regime, as shown in **Figure 5(c)**, the pressure on the contact area at the incompressible regime falls to the much lower "dynamic pressure" (Bernoulli stagnation pressure), which is a reasonable approximation verified by the experimental results of Soto et al. [52]. In addition to the water hammer pressure and dynamic pressure, three stress waves, that is, Rayleigh surface wave, compression wave and shear wave, are generated by the water hammer effect propagating on the target surface. Among these three stress waves, Rayleigh surface wave contains the most energy (i.e., over 2/3 of total energy) and has the lowest decay rate in terms of pressure magnitude so only it is considered in the further analysis of this study [53]. In isotropic solid-like superhydrophobic coatings, Rayleigh surface wave causes the surface molecules to move in ellipses in planes normal to the surface and parallel to the direction of propagation, which could lead to the crack initiation or surface topology change. As a result, Rayleigh surface wave can also cause the surface topology change and, consequently, induce wettability degradation over the SHS similar to water hammer pressure and dynamic pressure.

$$P_{hammer} = \rho_L c_L U_\infty \tag{4}$$

Eq. (4) is only applied to the ideal solid rigid surface. If the compressibility of the target surface is taken into consideration, the water hammer pressure could be expressed more accurately as:

$$P_{hammer} = \rho_L c_L U_\infty \left(\frac{1}{1+(\rho_L c_L / \rho_S c_S)} \right) \tag{5}$$

where ρ_s and c_s are the density and speed of sound in the target surface, respectively. The water hammer pressure is generated on a radius of contact given by $r_{hammer} = RU_\infty / c_L$ where R is the radius of a droplet before impact. The duration of water hammer pressure depends on the droplet radius as well as impact velocity and can be expressed as $t_{hammer} = 3RU_\infty / 2c_L^2$ [39].

As described by Blower [54] and Athanasopoulos et al. [55], the in-plane amplitude of the Rayleigh surface wave attenuates as it moves away from the impact center (i.e., $P \propto \xi^{-0.5}$ where P and ξ are the amplitude of Rayleigh surface wave and the traveling distance of it from the impact center, respectively) so the amplitude of Rayleigh surface wave generated by the water hammer pressure can be estimated by the following equation [26]:

$$P_{Rayleigh} = \rho_L c_L U_\infty \left(\frac{1}{1+(\rho_L c_L / \rho_S c_S)} \right) \left(\frac{r_{hammer}}{\xi} \right)^{0.5} \quad (6)$$

where ξ is the distance to the impact center.

All water hammer pressure, dynamic pressure and Rayleigh surface wave are likely to be the main factors causing wettability degradation after rain erosion tests since all of them have different features, which may generate severe damage to superhydrophobic coatings. Although water hammer pressure is tremendous in comparison with the dynamic pressure in terms of stress magnitude, the acting area of it is much smaller than that of the dynamic pressure. For a representative case of a droplet impinging onto an aluminum surface at a velocity of 65 m/s with a diameter of 3 mm, the magnitude of water hammer pressure can reach 89 MPa, significantly higher than only 2 MPa of dynamic pressure. However, the area experiencing the dynamic pressure is 512 times greater than the water hammer pressure. The significantly larger acting area of the dynamic pressure probably leads to more severe damage to the surface since more surface textures in a larger area might be damaged. The amplitude of the Rayleigh surface wave decays with propagating distance, which means the damage caused by the Rayleigh surface wave is higher near the impact center than in the far field. In summary, the importance of these three stresses to the degradation of the SHS due to rain erosion cannot be certainly compared with respect to the stress magnitude or acting area; thereby, the damage contributed by them should be further analyzed and compared considering energy transformation or classical surface fatigue theories.

To protect superhydrophobic coatings, previous studies have been devoted to fabricating durable coatings that resist droplets' high-speed impact. Brown et al. [56] applied plasma-enhanced chemical vapor deposition to fabricate a new durable superhydrophobic coating, which does not observe failure even after 6000 impacts at 165 m/s (typical speed at the tip of wind turbine blades). Song and Benmeddour [57] developed a new coating comprising a surface-modifying polymer bearing fluorinated and polydimethylsiloxane branches on the erosion-resistant polyurethane matrix to improve the droplet erosion durability. The development of protective coatings to resist droplet erosion should properly use several resins with different properties in combination or modifications to improve the superhydrophobicity as well as the durability [58].

3.2 Sand erosion

Sand erosion is a common and severe wear phenomenon for aircraft and wind turbines in the desert (see **Figure 6(a)**), which is caused by the repeated impact of sand or other abrasive particles [60]. During operation in desert, forests and near volcanoes, abrasive particles such as sand, dust, coal derivatives and fly ashes impinge on the surface at a high speed (over 100 m/s), leading to material removal and significant changes in the surface topography (see **Figure 6(b)**). Sand erosion occurs

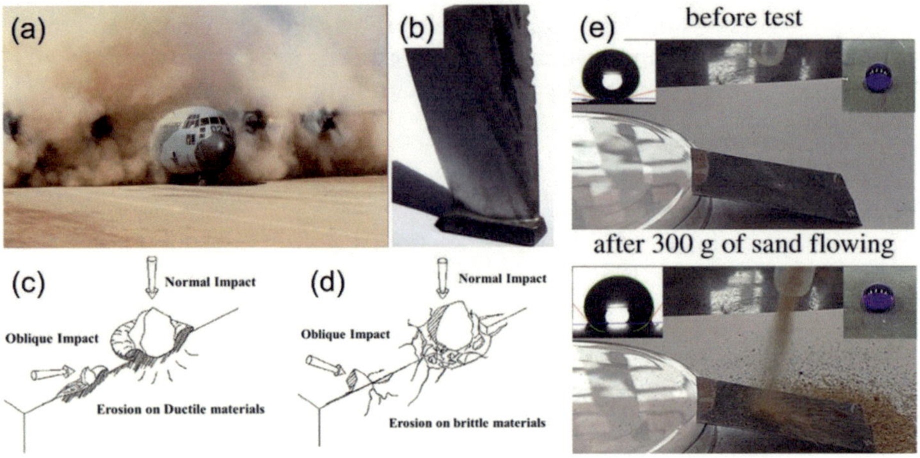

Figure 6.
Sand erosion is a severe problem to (a) aircraft and (b) turbines. The mechanism of sand erosion on (c) ductile and (d) brittle materials [59].

when the abrasive particles strike the coated surface at high velocities, causing the removal or thinning of the coating. While the material removal of ductile materials (e.g., metals and alloys) is caused by the cutting and plowing of the solid particles (see **Figure 6(c)**), sand erosion leading to crack initiation and propagation (see **Figure 6(d)**) on brittle materials (e.g., ceramics and silicon wafers) removes material pieces from the surface [61].

Many factors influence the material loss rate pertinent to sand erosion such as airflow speed, impact angle, object geometry, surface hardness, fracture toughness and material elasticity. Erosion rate can take the form:

$$W = CE_k^m f(\alpha) = CV_p^{2m} M_p^m f(\alpha) \tag{7}$$

where E_k, α, V_p and M_p are particle kinetic energy, impact angle, impact velocity and sand mass, respectively. C is a constant related to the material property. Particle kinetic energy has a form of $E_k^m = \frac{1}{2} M_p V_p^2$. C and m are constant terms related to the material characteristics, which need to be determined by experiments. While ultra-hard coatings are preferred to resist sand erosion, some softer polymers (e.g., polyurethane) can also provide excellent erosion resistance in atmospheric conditions [62]. Therefore, polyurethane has been used to fabricate durable superhydrophobic coatings [57].

Various erosion-resistant coatings for ice protection have been explored in pursuing durable superhydrophobic coatings. Kumar et al. [63] developed a superhydrophobic coating by introducing silica nanoparticles in a polyvinylidene fluoride matrix to improve erosion resistance. The sand erosion test was done by placing the sample surface 10 cm away from the sandblast nozzle at a pressure of 200 kPa for 30 s. After the sand erosion test, the coating with 10% polyvinylidene fluoride remains superhydrophobic. Fu et al. [64] fabricated an erosion-resistant superhydrophobic coating by applying plasma electrolytic oxidation (PEO) and chemical vapor deposition (CVD) on the aluminum substrate. After conducting a sand falling test (i.e., dropping sand particles from a certain height to the surface) using 2000 g sand, the coating remains superhydrophobic. Zheng et al. [65] employed a bio-epoxy resin and SiO_2 nanoparticles to make an eco-friendly hydrophobic coating. They found that the coating became superhydrophobic after conducting the sand erosion test since impacts of sand create a porous structure on the surface, allowing more air to be entrapped beneath the droplet.

3.3 Mechanical abrasion

When superhydrophobic coatings are subjected to mechanical abrasion, such as rubbing or scratching, the surface structures can be compromised, significantly degrading the superhydrophobic characteristics. Hensel et al. [66] proposed a method of creating superhydrophobic coatings by mimicking the effectively liquid-repellent and mechanically stable springtail-skin morphology. They tested the durability by oscillating a steel ball with a normal force on the bio-inspired coating. Hensel observed the failure of membranes, which are plastically deformed and delaminated from the substrate, and the pillar arrays that are broken from the bottom (see **Figure 7(a)**). Kondrashov et al. [67] presented an approach to generate a mechanically robust superhydrophobic coating by growing silicon microcones and nanograss using mask-free dry etching processes. They tested the coating durability by conducting a linear abrasion test, exposing the coating to a normal force and moving the coating back and forth. They found that the coating that combines microcone and nanograss structures (MC-LD) shows greatly improved mechanical durability since microcones can provide an excellent shield to resist wear damage (see **Figure 7(b)**). MC-LD remains superhydrophobic after the abrasion test with a force of 20 N, but the coating with only nanograss degrades to a hydrophilic coating after experiencing the same test. Dong et al. [68] used the 3D printing technique to fabricate a bulk superhydrophobic material by generating inherently nanoporous structures in the bulk material. In contrast to losing superhydrophobicity after 25 abrasion cycles for

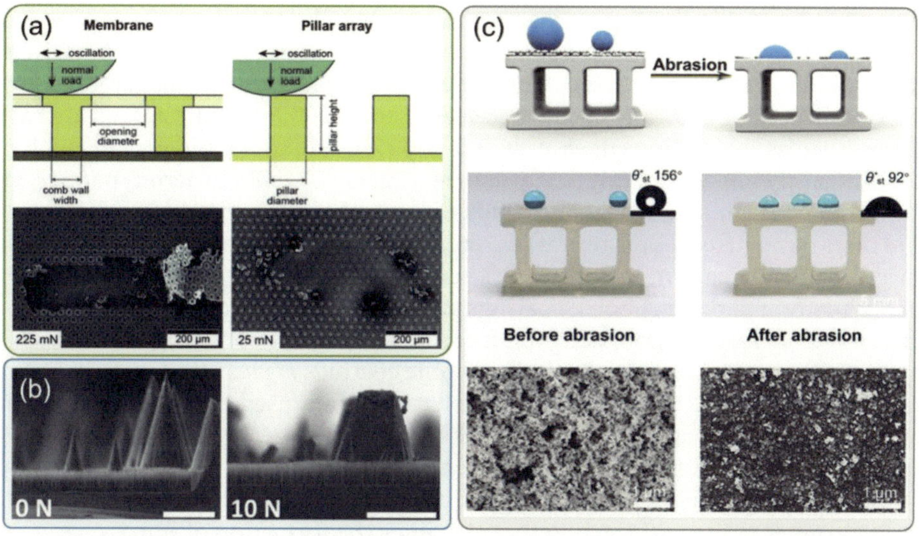

Figure 7.
Mechanical abrasion leads to the compromission of surface topology on superhydrophobic coatings.

those coatings (see **Figure** 7(**c**)), the bulk superhydrophobic material keeps low water adhesion and high contact angle after 40 abrasion cycles. Dong's research shows a promising method to fabricate durable superhydrophobic coatings without concerning the loss of materials.

The abrasion durability of superhydrophobic coatings can be tested using various methods. The most common method for measuring abrasion resistance involves linear shear abrasion, where a solid abradant moves tangentially across the surface being tested. In this type of wear, the contact point between the surfaces moves parallel to the surface with a force vector perpendicular to the material. Various experimental setups have been used in the literature to study this tangential abrasion, involving the fundamental concept of rubbing or sliding a solid material against the surface under investigation [69, 70]. This process causes mechanical wear, resulting in material removal and eventual alteration of the surface properties, including the loss of non-wettability. Circular and linear abrasion employs similar principles and utilizes the same experimental setup. The main distinction lies in the movement of abradant materials. Circular abrasion involves a disk-shaped abradant material that undergoes rotational motion. In contrast to linear abrasion, one drawback of circular abrasion is the variation in speeds among different sections of the abradant material. As a result, this type of testing is not as commonly used, though the setup is easier to implement. In certain situations, superhydrophobic coatings may encounter sharp objects such as knives, pens, forks, and other items that can cause significant surface damage. Previous studies have conducted blade tests for non-wettable coatings. These tests are essentially similar to linear abrasion tests, but a sharp blade simulates the rubbing action instead of using a flat abradant material.

4. Environmental factors affecting the durability of superhydrophobic coatings pertinent to aircraft/wind turbine icing mitigation

4.1 Icing/deicing cycles

Impact icing and bulk icing are two common icing phenomena related to ice accretion on the surface of aircraft. Impact icing occurs when supercooled large droplets or ice crystals in the atmosphere strike the surface of in-flight aircraft and freeze upon impact. Impact icing typically occurs between 0 and −20°C on unprotected aircraft surfaces, such as the leading edges of wings, engine guide vanes and propellers. It can significantly impact aircraft performance, causing changes in aerodynamics, and impaired control. Bulk icing usually occurs on parked aircraft in freezing drizzles where liquid water freezes on aircraft surfaces without impact or pressure. Bulk icing is smooth and transparent, making it difficult to detect visually.

Since many superhydrophobic coatings have excellent anti-icing performance, such as reducing ice adhesion, delaying freezing time and suppressing ice nucleation temperature, ice adhesion on superhydrophobic coatings is measured in plenty of studies. Bulk and impact ice are two common types of ice used in the measurement. The formation of impact ice is simulating the impact of supercooled droplets or ice crystals onto the aircraft surface using the experimental facility shown in **Figure 8(a)** [71]. Instead, some previous studies made impact ice by spraying droplets in a cold climate chamber to the surface directly [74]. Since the formation of impact ice involves the impact of droplets, the surface topology of superhydrophobic coatings could be compromised in the same mechanism as the droplet erosion in this process.

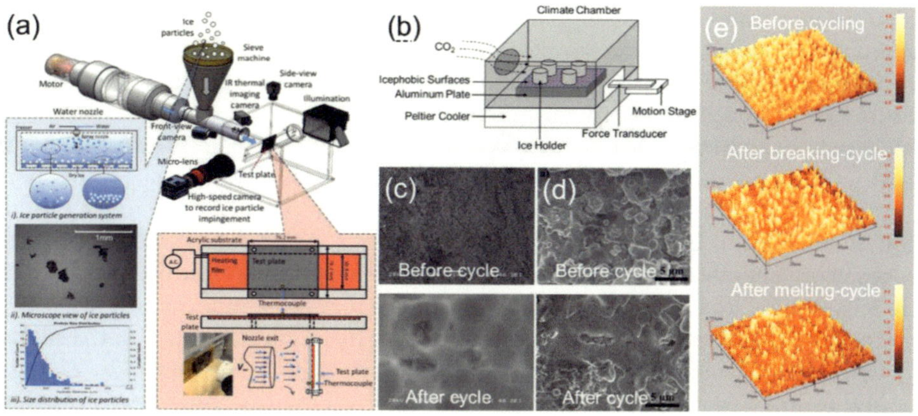

Figure 8.
The experimental setup of creating (a) impact ice samples [71] and (b) bulk ice samples [42]. (c–e) Icing-deicing cycles significantly change the surface topology of superhydrophobic coatings [56, 72, 73].

Besides, before ice nucleation of droplets on the surface, water can infiltrate micro-/nanostructures of superhydrophobic coatings. When water freezes and expands, it can potentially damage or break the surface topology. While impact ice can simulate the in-flight icing process more closely and evaluate the icephobic performance of a coating more accurately, bulk ice is used more frequently due to simple experimental conditions. **Figure 8(b)** shows a typical facility for icing/deicing cycles that create the bulk ice at first and break the ice sample mechanically then [42]. Another icing/deicing cycle is creating the bulk ice also at first but melting the ice consequently. Although melting bulk ice seems harmless to the superhydrophobic coating, removing the surface material or applying tremendous fracture stress in the icing process could significantly change the surface topology. Mobarakeh and Farzaneh [72] observed that the spongy micro-structures on the surface is replaced by large roughness features after 15 icing/deicing cycles (see **Figure 8(c)**). Mobarakeh and Farzaneh conduced the icing test at −10°C and the deicing test at the room temperature, which are the icing/deicing conditions used in many literature. Their icing samples were obtained in an icing wind tunnel to simulate the impact icing on UAV. Brown et al. [56] reported that the amount of material is significantly reduced after 80 cycles (See **Figure 8(d)**). Similarly, Wang et al. [73] proposed that the amount of microcones on the superhydrophobic coating is significantly reduced after 40 melting cycles (See **Figure 8(e)**). They also conducted the ice-breaking tests using the facility shown in **Figure 8(b)**, and surprising found that 20 breaking cycles contribute less damage to the superhydrophobic coating than 40 melting cycles in terms of surface roughness and microcone number. Their study indicates that the water freezing process severely damages the hierarchical structures more than removing ice from the surface.

4.2 Chemical corrosion

Corrosion is usually an undesirable and unsecured phenomenon since it has a detrimental impact on the functional characteristics of metals. The total annual cost of corrosion in the petroleum industry of USA is estimated to be 1.4 billion dollars [75]. Superhydrophobic coatings have shown excellent corrosion resistance in highly acidic or aggressive media [76]. The fabricated hierarchical structures offer a passivation layer hindering metallic corrosion, e.g., offshore wind turbines subjected to aggressive environments with high humidity and intensive salinity, offering a new efficient mechanism from anti-corrosion. Liu et al. [77] stated that the anti-corrosion mechanism of superhydrophobic coatings under salty water is avoiding corrosive medium like Cl^- reaching the metal substrate.

While superhydrophobic coatings have excellent anti-corrosion features, chemical corrosion could also lead to the degradation of hydrophobicity. An important chemical corrosion (or called contamination) subjected to aircraft ground icing is the corrosion of deicing fluids to superhydrophobic coatings. Aircraft ground deicing often involves using chemicals such as deicers or anti-icing agents. While these chemicals aid in ice removal, they can also degrade the performance of superhydrophobic coating by resolving the coating material or depositing hydrophilic agents on the surface since deicing fluids usually contain glycols and other organic solvents [42]. Zhang et al. [42] found that a commercial type I deicing fluid mainly containing propylene glycol could degrade a silicon-based superhydrophobicity by resolving the topping material and significantly undermining the hierarchical structure. Furthermore, the thickening agents (e.g., xanthan gum) in a type IV deicing fluid could adhere tightly on superhydrophobic coatings, which results in the transition of surface wettability from superhydrophobic to hydrophilic very quickly [42]. Even though the coating intrinsic chemical characteristics are not altered in this case, the surface is covered by another hydrophilic material in the aircraft deicing process, which should be cautious for the application of superhydrophobic coatings on aircraft.

4.3 UV irradiation

In the presence of oxygen or water in the environment, UV in sunlight could induce the photo-oxidation of coatings. The UV resistance of superhydrophobic coatings varies with the coating composition. Shorter wavelengths in the UV range result in more severe damage to the opaque coating, whereas UV-transparent coatings can very well resist the photo-oxidation induced by UV light. Superhydrophobic coatings that only absorb UV light with wavelengths lower than 280 nm usually have good durability to resist UV radiation [33].

The UV durability of superhydrophobic coatings could be assessed by placing the coatings under sunlight in order to cover the whole range of sunlight spectrum but this method is not time efficient [78]. A climate chamber equipped with an intensive UV light lamp can accelerate the degradation of superhydrophobic coating [79]. An international standard, that is, ASTM G154, has regulated the standard practice for operating UV lamp for exposure of non-metallic materials, which is widely used in many previous studies of testing UV degradation of superhydrophobic coatings [80, 81].

Plenty of studies are dedicated to fabricating UV-resistant superhydrophobic coatings. Radwan et al. [82] combined poly(vinylidene fluoride-co-hexafluoropropylene) with alumina nanoparticles to create a superhydrophobic coating with high UV durability. Their results showed that after a long 500-hour UV exposure at a constant temperature of 60°C, the coating remained superhydrophobic with little contact angle degradation. They contribute excellent UV durability to the higher bond energy provided by alumina nanoparticles. Raghu et al. [83] employed transparent ZrO_2-nanotubes to fabricate transparent superhydrophobic coatings to improve UV-resistance. Gao et al. [84] developed a highly transparent superhydrophobic coating with SiO_2-coated ZnO nanorod arrays. A thin layer of SiO_2 acts as a physical barrier to filter UV to protect ZnO, which is typically prone to UV-triggered photo-oxidation reactions.

Figure 9.
Insect contamination on an (a) aircraft and a (b) wind turbine. (c) Insect-contaminated wind turbine generates less power than the clean turbine [85].

4.4 Insect contamination

Insect impact are operational events of aircraft occurring during the ground operation, take-off and landing phases of the flight profile, which was recognized as early as 1945. Insect impact residue can greatly contaminate the aircraft surface (see **Figure 9(a)**) and alter the aerodynamic performance of aircraft [86]. The issue of insect debris has garnered attention in multiple industries. Specifically, it has been a focus for the wind farm industry. There has been a particular interest in understanding the effect of surface irregularities induced by the insect contamination (see **Figure 9(b)**) on the efficiency of wind turbine blades and the subsequent power losses they may cause. Corten and Veldkamp [85] hypothesized that the insect debris at the suction peak of turbine blades could reduce the power output and verified this hypothesis experimentally (see **Figure 9(c)**).

Superhydrophobic coatings are effective in reducing insect contamination by significantly decreasing the adhesion of the debris. This property makes it difficult for insects to adhere to the coating. Krishnan et al. [87] observed that the superhydrophobic coating could reduce the contamination area efficiently since the superhydrophobic coating significantly promotes the retracting of insect debris. However, they also pointed out that superhydrophobic and oleophobic coatings are ideal for minimizing insect contamination because the adhesion of insect blood is low on oleophobic coatings rather than superhydrophobic coatings. Kok et al. [88] tested the ability of contamination cleaning for surfaces with various roughness and proposed that rougher surfaces with better hydrophobicity display less residue adhesion.

5. Recommendations for the durability of superhydrophobic coatings pertinent to aircraft/wind turbine icing mitigation

This chapter reviews the recent advances in durability of superhydrophobic coatings pertinent to aircraft and wind turbine icing mitigation. In order to develop effective superhydrophobic coating for aircraft/wind turbines, it is necessary to have a consistent, reliable method for characterizing the performance of candidate materials subjected to realistic environmental conditions, such as rain, sand and ultraviolet exposure. As a result, the durability tests listed above (i.e., droplet erosion resistance, sand erosion resistance, abrasion resistance, icing/deicing resistance, UV resistance, chemical corrosion resistance and insect contamination) are necessary to be conducted to quantify the durability of superhydrophobic coatings before applying them on aircraft and wind turbines in practice. This chapter is intended to display the problem and perhaps can shed some light into conducting tests that can be industrially valuable. Here are some recommendations for the durability test of superhydrophobic coatings:

1. In order to assist the advances in the application of superhydrophobic coatings, it is significant to standardize the evaluation methods based on the working conditions of aircraft and wind turbines. This chapter indicates that some of the above durability testing methods are well studied in this regard, and this allows specific recommendations. **Table 1** summarizes the standard methods for durability assessment of a coating. However, there is currently a lack of specific durability testing standards for superhydrophobic coatings that can be applied to aircraft or wind turbines. As a result, some standards or regulations for coating durability assessment should be developed or updated based on current standards. For example, ASTM G73 regulates the droplet erosion durability of coatings by quantifying the material removal rate. However, superhydrophobic coatings suffer degradation (e.g., a decrease of the static contact angle) after the droplet erosion test with little material removal. In this way, ASTM G73 should be updated to include the hydrophobicity degradation after the droplet erosion test to meet the specific requirement of applications of superhydrophobic coatings on aircraft.

2. The importance of each durability test needs to be quantified based on local environmental conditions since natural factors that influence the durability of superhydrophobic coating vary with different regions. For example, sand erosion is severe in Middle Eastern countries where the desert covers a large portion of countries, but it is not a primary concern in countries in Western European countries that receive a significant amount of rainfall throughout the year from the Atlantic Ocean. Instead, superhydrophobic coatings prepared for offshore wind turbines in Europe are severely threatened by the corrosion of salty vapor and droplet erosion due to plenty of airborne tiny droplets.

3. The durability tests of superhydrophobic coatings should be application specific. Fabricating a superhydrophobic, durable coating for all tests listed above

Durability type	Standard code	Test details
Droplet erosion	ASTM G73	Samples are installed on a rotating arm and impinge with droplets at high speed (up to 600 m/s).
Sand erosion	ASTM G76	Samples are eroded by high-speed airborne particles injected by a nozzle with a specified particle size, velocity, and attack angle.
Mechanical abrasion	ASTM D4060	Abrasion of samples with different loads and abraders.
Chemical corrosion	ASTM G1654	Evaluating basic corrosion performance of coatings after exposure to corrosive environments.
UV	ASTM G154	Simulation of weathering conditions in a chamber under consecutive UV irradiation.
Icing/deicing cycles	ASTM C1645	Determining the effects of icing/deicing cycles on the length of service of concrete paving units.

Table 1.
ASTM standards about coating durability.

is an overwhelming task. Despite this fact, the coating should also be excellent if it is durable under the situations where it mostly works. Offshore wind turbines mostly work in aggressive environments with high humidity, severe corrosion and intense UV radiation but sand erosion is rare. As a result, developing a superhydrophobic coating with excellent UV and corrosion resistance is an ideal coating for the icing mitigation of offshore wind turbines. Instead, aircraft working in airports in the desert faces intense UV radiation in park and severe sand erosion during landing and take-off, so a superhydrophobic coating with excellent erosion and UV resistance would be suitable for aircraft icing mitigation.

4. Coating durability needs to be further improved to allow the coating to perform well in operating conditions of aircraft and wind turbines. The coating should be tested in a realistic and dynamic environment, such as inside a droplet erosion test rig where the droplet can impinge onto the coating with a speed over 100 m/s analogous to the landing or take-off condition.

6. Conclusions

In conclusion, superhydrophobic coatings that can create large microcracks have demonstrated significant potential in anti-icing applications. These coatings exhibit excellent water repellency, creating a protective barrier on surfaces that prevent ice formation. Superhydrophobic coatings can effectively reduce ice accumulation and enhance anti-icing properties by minimizing the adhesion of water droplets/ice. Compared to conventional thermal-based anti-icing technologies, superhydrophobic coatings have advantages of lightweight and low energy consumption. As a result, applying superhydrophobic coatings to achieve surface modification is a promising anti-icing technique in wide applications including aerospace and wind energy. International Organization for Standardization (ISO) has published a standard (ISO10689) to systematically evaluate the durability of superhydrophobic coatings pertinent to industrial applications.

While superhydrophobic coatings show promising results in anti-icing applications, further research is still needed to optimize their performance and extend the length of service. This chapter mainly reviews the challenges in improving the durability of superhydrophobic coatings and highlights recent advances in enhancing the coating durability. One of the challenges is droplet erosion, which could occur on in-flight aircrafts or offshore wind turbines subject to the high-speed impact of droplets. The “water hammer effect” in the high-speed impact generates tremendous stress, leading to severe coating damage. Further, our recent study indicates that the water hammer pressure and Rayleigh surface wave induced by the impact are the main factors leading to the coating degradation in droplet erosion. Sand erosion, mechanical abrasion and icing/deicing test also significantly change the surface topology, but UV radiation and chemical corrosion change the surface chemistry. In this way, the durability of superhydrophobic coatings with hierarchical structures can be defined as the ability of a surface to maintain surface topology and low surface energy under practical working conditions. As a result, durable superhydrophobic coatings should be fabricated with excellent mechanical hardness and stable chemical properties.

Acknowledgements

The author would like to thank the support of Zhongfa Aviation Institute of Beihang University.

Author details

Zichen Zhang[1*], Junfei Ou[2] and Wen Li[2]

1 Zhongfa Aviation Institute, Beihang University, Pingyao Town, China

2 Jiangsu University of Technology, Changzhou, China

*Address all correspondence to: zichenzhang437@gmail.com

References

[1] Cao Y, Tan W, Wu Z. Aircraft icing: An ongoing threat to aviation safety. Aerospace Science and Technology. 2018;**75**:353-385. DOI: 10.1016/j.ast.2017.12.028

[2] De Pauw D, Dolatabadi A. Effect of superhydrophobic coating on the anti-icing and deicing of an airfoil. Journal of Aircraft. 2017;**54**:490-499. DOI: 10.2514/1.C033828

[3] Han N, Siddique MA, Zhang Z, Tian L, Hu H, Hu H. A flight-testing campaign to examine inflight icing characteristics and its effects on the flight performance of an unmanned-aerial-vehicle. Cold Regions Science and Technology. 2023;**207**:103775. DOI: 10.1016/j.coldregions.2023.103775

[4] Botura G, Fahrner A. Icing Detection System–Conception, Development, Testing and Applicability to UAVs. San Diego, California: Goodrich Corp AIAA; 2003

[5] Zhang B, Tang L, Roemer M. Probabilistic weather forecasting analysis for unmanned aerial vehicle path planning. Journal of Guidance, Control, and Dynamics. 2014;**37**:309-312. DOI: 10.2514/1.61651

[6] Li L, Liu Y, Tian L, Hu H, Hu H, Liu X, et al. An experimental study on a hot-air-based anti- /de-icing system for aero-engine inlet guide vanes. Applied Thermal Engineering. 2020;**167**:114778. DOI: 10.1016/j.applthermaleng.2019.114778

[7] Liu Y, Kolbakir C, Hu H, Meng X, Hu H. An experimental study on the thermal effects of duty-cycled plasma actuation pertinent to aircraft icing mitigation. International Journal of Heat and Mass Transfer. 2019;**136**:864-876. DOI: 10.1016/j.ijheatmasstransfer.2019.03.068

[8] Muthumani A, Fay L, Akin M, Wang S, Gong J, Shi X. Correlating lab and field tests for evaluation of deicing and anti-icing chemicals: A review of potential approaches. Cold Regions Science and Technology. 2014;**97**:21-32. DOI: 10.1016/j.coldregions.2013.10.001

[9] Gao L, Liu Y, Ma L, Hu H. A hybrid strategy combining minimized leading-edge electric-heating and superhydro-/ice-phobic surface coating for wind turbine icing mitigation. Renewable Energy. 2019;**140**:943-956. DOI: 10.1016/j.renene.2019.03.112

[10] Huang X, Tepylo N, Pommier-Budinger V, Budinger M, Bonaccurso E, Villedieu P, et al. A survey of icephobic coatings and their potential use in a hybrid coating/active ice protection system for aerospace applications. Progress in Aerospace Science. 2019;**105**:74-97. DOI: 10.1016/j.paerosci.2019.01.002

[11] Cohen N, Dotan A, Dodiuk H, Kenig S. Superhydrophobic coatings and their durability. Materials and Manufacturing Processes. 2016;**31**:1143-1155. DOI: 10.1080/10426914.2015.1090600

[12] Hooda A, Goyat MS, Pandey JK, Kumar A, Gupta R. A review on fundamentals, constraints and fabrication techniques of superhydrophobic coatings. Progress in Organic Coatings. 2020;**142**:105557. DOI: 10.1016/j.porgcoat.2020.105557

[13] Yang N, Li J, Bai N, Xu L, Li Q. One step phase separation process to fabricate superhydrophobic PVC films

and its corrosion prevention for AZ91D magnesium alloy. Materials Science and Engineering B: Solid-State Materials for Advanced Technology. 2016;**209**:1-9. DOI: 10.1016/j.mseb.2016.01.009

[14] Ding B, Ogawa T, Kim J, Fujimoto K, Shiratori S. Fabrication of a super-hydrophobic nanofibrous zinc oxide film surface by electrospinning. Thin Solid Films. 2008;**516**:2495-2501. DOI: 10.1016/j.tsf.2007.04.086

[15] Yang Y, He H, Li Y, Qiu J. Using nanoimprint lithography to create robust, buoyant, superhydrophobic PVB/SiO2 coatings on wood surfaces inspired by red roses petal. Scientific Reports. 2019;**9**:1-9. DOI: 10.1038/s41598-019-46337-y

[16] Zhu X, Zhang Z, Ren G, Yang J, Wang K, Xu X, et al. A novel superhydrophobic bulk material. Journal of Materials Chemistry. 2012;**22**:20146-20148. DOI: 10.1039/c2jm33769j

[17] Singh AK, Singh JK. Fabrication of durable superhydrophobic coatings on cotton fabrics with photocatalytic activity by fluorine-free chemical modification for dual-functional water purification. New Journal of Chemistry. 2017;**41**:4618-4628. DOI: 10.1039/c7nj01042g

[18] Wang D, Sun Q, Hokkanen MJ, Zhang C, Lin FY, Liu Q, et al. Design of robust superhydrophobic surfaces. Nature. 2020;**582**:55-59. DOI: 10.1038/s41586-020-2331-8

[19] Li L, Roethel S, Breedveld V, Hess DW. Creation of low hysteresis superhydrophobic paper by deposition of hydrophilic diamond-like carbon films. Cellulose. 2013;**20**:3219-3226. DOI: 10.1007/s10570-013-0078-1

[20] Leary WM. We Freeze to Please: A History of NASA's Icing Research Tunnel and the Quest for Flight Safety. Vol. 4226. National Aeronautics and Space Administration; 2002

[21] Struk P, Broeren A, Potapczuk M, Kreeger RE, Blankenship K, Chen R-C, et al. NASA Analysis of Alternatives Study for Icing Research. Cleveland, Ohio, USA: National Aeronautics and Space Administration; 2022

[22] Sojoudi H, Wang M, Boscher ND, McKinley GH, Gleason KK. Durable and scalable icephobic surfaces: Similarities and distinctions from superhydrophobic surfaces. Soft Matter. 2016;**12**:1938-1963. DOI: 10.1039/c5sm02295a

[23] Milionis A, Loth E, Bayer IS. Recent advances in the mechanical durability of superhydrophobic materials. Advances in Colloid and Interface Science. 2016;**229**:57-79. DOI: 10.1016/j.cis.2015.12.007

[24] Gorb SN, Gorb EV. Anti-icing strategies of plant surfaces: The ice formation on leaves visualized by cryo-SEM experiments. Science of Nature. 2022;**109**:24. DOI: 10.1007/s00114-022-01789-7

[25] Byun S, Jeong H, Kim DR, Lee KS. Frost layer growth behavior on ultra-low temperature surface with a superhydrophobic coating. International Communications in Heat and Mass Transfer. 2021;**128**:105641. DOI: 10.1016/j.icheatmasstransfer.2021.105641

[26] Zhang Z, Ma L, Liu Y, Ren J, Hu H. An experimental study of rain erosion effects on a hydro-/ice-phobic coating pertinent to unmanned-arial-system (UAS) inflight icing mitigation. Cold Regions Science and Technology. 2021;**181**:103196. DOI: 10.1016/j.coldregions.2020.103196

[27] Gohardani O. Impact of erosion testing aspects on current and future

flight conditions. Progress in Aerospace Science. 2011;**47**:280-303. DOI: 10.1016/j.paerosci.2011.04.001

[28] Tian X, Verho T, Ras RHA. Moving superhydrophobic surfaces toward real-world applications. Science (80-). 2016;**352**:142-143. DOI: 10.1126/science.aaf2073

[29] Wang L, Tian Z, Jiang G, Luo X, Chen C, Hu X, et al. Spontaneous dewetting transitions of droplets during icing & melting cycle. Nature Communications. 2022;**13**:378. DOI: 10.1038/s41467-022-28036-x

[30] Ma L, Zhang Z, Gao L, Liu Y, Hu H. An exploratory study on using slippery-liquid-infused-porous-surface (SLIPS) for wind turbine icing mitigation. Renewable Energy. 2020;**162**:2344-2360. DOI: 10.1016/j.renene.2020.10.013

[31] Xiu Y, Liu Y, Hess DW, Wong CP. Mechanically robust superhyd rophobicity on hierarchically structured Si surfaces. Nanotechnology. 2010;**21**:155705. DOI: 10.1088/0957-4484/21/15/155705

[32] Myronyuk O, Baklan D, Yong Z, Rodin AM. Complex destruction of textured water-repellent coatings under the influence of UV and water flow. Materials Today Communications. 2022;**33**:104509. DOI: 10.1016/j.mtcomm.2022.104509

[33] Hakimian A, Nazifi S, Ghasemi H. Durability assessment of icephobic coatings. In: Mittal KL, Choi CH, editors. Ice Adhesion. 2020. DOI: 10.1002/9781119640523.ch17

[34] Dalawai SP, Saad Aly MA, Latthe SS, Xing R, Sutar RS, Nagappan S, et al. Recent advances in durability of superhydrophobic self-cleaning technology: A critical review. Progress in Organic Coatings. 2020;**138**:105381. DOI: 10.1016/j.porgcoat.2019.105381

[35] Chen H, Muros-Cobos JL, Amirfazli A. Contact angle measurement with a smartphone. The Review of Scientific Instruments. 2018;**89**(3):035117. DOI: 10.1063/1.5022370

[36] Huang X, Gates I. Apparent contact angle around the periphery of a liquid drop on roughened surfaces. Scientific Reports. 2020;**10**:8220. DOI: 10.1038/s41598-020-65122-w

[37] Schrader ME. Young-dupre revisited. Langmuir. 1995;**11**:3585-3589. DOI: 10.1021/la00009a049

[38] Yang C, Zeng Q, Huang J, Guo Z. Droplet manipulation on superhydrophobic surfaces based on external stimulation: A review. Advances in Colloid and Interface Science. 2022;**306**:102724. DOI: 10.1016/j.cis.2022.102724

[39] Work A, Lian Y. A critical review of the measurement of ice adhesion to solid substrates. Progress in Aerospace Science. 2018;**98**:1-26. DOI: 10.1016/j.paerosci.2018.03.001

[40] Meuler AJ, Smith JD, Varanasi KK, Mabry JM, McKinley GH, Cohen RE. Relationships between water wettability and ice adhesion. ACS Applied Materials & Interfaces. 2010;**2**:3100-3110. DOI: 10.1021/am1006035

[41] Butt HJ, Liu J, Koynov K, Straub B, Hinduja C, Roismann I, et al. Contact angle hysteresis. Current Opinion in Colloid & Interface Science. 2022;**59**:101574. DOI: 10.1016/j.cocis.2022.101574

[42] Zhang Z, Lusi A, Hu H, Bai X, Hu H. An experimental study on the detrimental effects of deicing fluids on

the performance of icephobic coatings for aircraft icing mitigation. Aerospace Science and Technology. 2021;**119**:107090. DOI: 10.1016/j.ast.2021.107090

[43] Zhang Z, Ou J, Li W, Amirfazli A. Folding characteristics of membranes in capillary origami. Journal of Colloid and Interface Science. 2023;**630**:111-120. DOI: 10.1016/j.jcis.2022.10.046

[44] Bengaluru Subramanyam S, Kondrashov V, Rühe J, Varanasi KK. Low ice adhesion on nano-textured superhydrophobic surfaces under supersaturated conditions. ACS Applied Materials & Interfaces. 2016;**8**:12583-12587. DOI: 10.1021/acsami.6b01133

[45] Herring R, Dyer K, Martin F, Ward C. The increasing importance of leading edge erosion and a review of existing protection solutions. Renewable and Sustainable Energy Reviews. 2019;**115**:109382. DOI: 10.1016/j.rser.2019.109382

[46] Keegan MH, Nash DH, Stack MM. On erosion issues associated with the leading edge of wind turbine blades. Journal of Physics D: Applied Physics. 2013;**46**:383001. DOI: 10.1088/0022-3727/46/38/383001

[47] Zhan Y, Yu S, Amirfazli A, Rahim Siddiqui A, Li W. Recent advances in antibacterial superhydrophobic coatings. Advanced Engineering Materials. 2022;**24**:1-22. DOI: 10.1002/adem.202101053

[48] Fujisawa N, Takano S, Fujisawa K, Yamagata T. Experiments on liquid droplet impingement erosion on a rough surface. Wear. 2018;**398-399**:158-164. DOI: 10.1016/j.wear.2017.12.003

[49] Ghidaoui MS, Zhao M, McInnis DA, Axworthy DH. A review of water hammer theory and practice. Applied Mechanics Reviews. 2005;**58**:49. DOI: 10.1115/1.1828050

[50] Field JE. The physics of liquid impact, shock wave interactions with cavities, and the implications to shock wave lithotripsy. Physics in Medicine and Biology. 1991;**36**:1475-1484. DOI: 10.1088/0031-9155/36/11/007

[51] Lesser MB. Analytic solutions of liquid-drop impact problems. Proceedings of the Royal Society A: Mathematical, Physical and Engineering Sciences. 1981;**377**:289-308. DOI: 10.1098/rspa.1981.0125

[52] Soto D, De Larivière AB, Boutillon X, Clanet C, Quéré D. The force of impacting rain. Soft Matter. 2014;**10**:4929-4934. DOI: 10.1039/C4SM00513A

[53] Grundwürmer M, Nuyken O, Meyer M, Wehr J, Schupp N. Sol-gel derived erosion protection coatings against damage caused by liquid impact. Wear. 2007;**263**:318-329. DOI: 10.1016/j.wear.2006.12.039

[54] Blowers RM. On the response of an elastic solid to droplet impact. IMA Journal of Applied Mathematics. 1969;**5**:167-193. DOI: 10.1093/imamat/5.2.167

[55] Athanasopoulos GA, Pelekis PC, Anagnostopoulos GA. Effect of soil stiffness in the attenuation of Rayleigh-wave motions from field measurements. Soil Dynamics and Earthquake Engineering. 2000;**19**:277-288. DOI: 10.1016/S0267-7261(00)00009-9

[56] Brown S, Lengaigne J, Sharifi N, Pugh M, Moreau C, Dolatabadi A, et al. Durability of superhydrophobic duplex coating systems for aerospace applications. Surface and Coatings Technology. 2020;**401**:126249. DOI: 10.1016/j.surfcoat.2020.126249

[57] Song N, Benmeddour A. Erosion resistant hydrophobic coatings for

passive ice protection of aircraft. Applied Sciences. 2022;**12**:9589. DOI: 10.3390/app12199589

[58] Chen J, Wang J, Ni A. A review on rain erosion protection of wind turbine blades. Journal of Coatings Technology and Research. 2019;**16**:15-24. DOI: 10.1007/s11998-018-0134-8

[59] Meena MK, Tudu BK, Kumar A, Bhushan B. Development of polyurethane-based superhydrophobic coatings on steel surfaces. Philosophical Transactions of the Royal Society A—Mathematical Physical and Engineering Sciences. 2020;**378**:20190446. DOI: 10.1098/rsta.2019.0446

[60] Hutchings IM, Winter RE. Particle erosion of ductile metals: A mechanism of material removal. Wear. 1974;**27**:121-128. DOI: 10.1016/0043-1648(74)90091-X

[61] Wang YF, Yang ZG. Finite element model of erosive wear on ductile and brittle materials. Wear. 2008;**265**:871-878. DOI: 10.1016/j.wear.2008.01.014

[62] Wood RJK. The sand erosion performance of coatings. Materials and Design. 1999;**20**:179-191. DOI: 10.1016/s0261-3069(99)00024-2

[63] Kumar D, Li L, Chen Z. Mechanically robust polyvinylidene fluoride (PVDF) based superhydrophobic coatings for self-cleaning applications. Progress in Organic Coatings. 2016;**101**:385-390. DOI: 10.1016/j.porgcoat.2016.09.003

[64] Fu J, Sun Y, Ji Y, Zhang J. Fabrication of robust ceramic based superhydrophobic coating on aluminum substrate via plasma electrolytic oxidation and chemical vapor deposition methods. Journal of Materials Processing Technology. 2022;**306**:117641. DOI: 10.1016/j.jmatprotec.2022.117641

[65] Zheng S, Bellido-Aguilar DA, Huang Y, Zeng X, Zhang Q, Chen Z. Mechanically robust hydrophobic bio-based epoxy coatings for anti-corrosion application. Surface and Coatings Technology. 2019;**363**:43-50. DOI: 10.1016/j.surfcoat.2019.02.020

[66] Hensel R, Finn A, Helbig R, Braun HG, Neinhuis C, Fischer WJ, et al. Biologically inspired omniphobic surfaces by reverse imprint lithography. Advanced Materials. 2014;**26**:2029-2033. DOI: 10.1002/adma.201305408

[67] Kondrashov V, Rühe J. Microcones and nanograss: Toward mechanically robust superhydrophobic surfaces. Langmuir. 2014;**30**:4342-4350. DOI: 10.1021/la500395e

[68] Dong Z, Vuckovac M, Cui W, Zhou Q, Ras RHA, Levkin PA. 3D printing of superhydrophobic objects with bulk nanostructure. Advanced Materials. 2021;**33**:2106068. DOI: 10.1002/adma.202106068

[69] Huang J, Li M, Ren C, Huang W, Miao Y, Wu Q, et al. Construction of HLNPs/Fe3O4based superhydrophobic coating with excellent abrasion resistance, UV resistance, flame retardation and oil absorbency. Journal of Environmental Chemical Engineering. 2023;**11**:109046. DOI: 10.1016/j.jece.2022.109046

[70] Dong S, Yang Y, Ma R, Du A, Yang M, Fan Y, et al. Investigation of the anti-corrosion performance of a Cu/Ni–B4C superhydrophobic coating with self-cleaning and abrasion resistance properties. Surface and Coatings Technology. 2022;**441**:128568. DOI: 10.1016/j.surfcoat.2022.128568

[71] Hu H, Tian L, Hu H. Experimental investigation on ice accretion upon ice particle impacting onto heated surface. AIAA Journal. 2023;**61**:3019-3031. DOI: 10.2514/1.j062425

[72] Foroughi Mobarakeh L, Jafari R, Farzaneh M. The ice repellency of plasma polymerized hexamethyldisiloxane coating. Applied Surface Science. 2013;**284**:459-463. DOI: 10.1016/j.apsusc.2013.07.119

[73] Wang Y, Xue J, Wang Q, Chen Q, Ding J. Verification of icephobic/anti-icing properties of a superhydrophobic surface. ACS Applied Materials & Interfaces. 2013;**5**:3370-3381. DOI: 10.1021/am400429q

[74] Tetteh E, Loth E. Reducing static and impact ice adhesion with a self-lubricating icephobic coating (SLIC). Coatings. 2020;**10**:262. DOI: 10.3390/coatings10030262

[75] Wang ZM, Song GL, Zhang J. Corrosion control in CO2 enhanced oil recovery from a perspective of multiphase fluids. Frontiers in Materials. 2019;**6**:1-16. DOI: 10.3389/fmats.2019.00272

[76] Gupta R, Verma R, Kango S, Constantin A, Kharia P, Saini R, et al. A critical review on recent progress, open challenges, and applications of corrosion-resistant superhydrophobic coating. Materials Today Communications. 2023;**34**:105201. DOI: 10.1016/j.mtcomm.2022.105201

[77] Liu T, Yin Y, Dong L. New application of the "underwater super-hydrophobic" surface in the corrosion protection. Advances in Materials Research. 2009;**79-82**:1115-1118. DOI: 10.4028/www.scientific.net/AMR.79-82.1115

[78] Irajizad P, Nazifi S, Ghasemi H. Icephobic surfaces: Definition and figures of merit. Advances in Colloid and Interface Science. 2019;**269**:203-218. DOI: 10.1016/j.cis.2019.04.005

[79] Sataeva NE, Boinovich LB, Emelyanenko KA, Domantovsky AG, Emelyanenko AM. Laser-assisted processing of aluminum alloy for the fabrication of superhydrophobic coatings withstanding multiple degradation factors. Surface and Coatings Technology. 2020;**397**:125993. DOI: 10.1016/j.surfcoat.2020.125993

[80] Allahdini A, Momen G, Munger F, Brettschneider S, Fofana I, Jafari R. Performance of a nanotextured superhydrophobic coating developed for high-voltage outdoor porcelain insulators. Colloids and Surfaces A: Physicochemical and Engineering Aspects. 2022;**649**:129461. DOI: 10.1016/j.colsurfa.2022.129461

[81] Foroughi Mobarakeh L, Jafari R, Farzaneh M. Robust icephobic, and anticorrosive plasma polymer coating. Cold Regions Science and Technology. 2018;**151**:89-93. DOI: 10.1016/j.coldregions.2018.03.009

[82] Radwan AB, El-Hout SI, Ibrahim MAM, Ismail EH, Abdullah AM. Superior corrosion and UV-resistant highly porous poly(vinylidene fluoride-co-hexafluoropropylene)/alumina superhydrophobic coating. ACS Applied Polymer Materials. 2022;**4**:1358-1367. DOI: 10.1021/acsapm.1c01710

[83] Raghu SNV, Chuluunbandi K, Killian MS. Zirconia nanotube coatings—UV-resistant superhydrophobic surfaces. Surfaces and Interfaces. 2021;**26**:101357. DOI: 10.1016/j.surfin.2021.101357

[84] Gao Y, Gereige I, El Labban A, Cha D, Isimjan TT, Beaujuge PM. Highly transparent and UV-resistant superhydrophobic SiO2-coated ZnO nanorod arrays. ACS Applied Materials & Interfaces. 2014;**6**:2219-2223. DOI: 10.1021/am405513k

[85] Corten GP, Veldkamp HF. Aerodynamics: Insects can halve wind-turbine power. Nature. 2001;**412**:41-42. DOI: 10.1038/35083698

[86] Kok M, Smith JG, Wohl CJ, Siochi EJ, Young TM. Critical considerations in the mitigation of insect residue contamination on aircraft surfaces—A review. Progress in Aerospace Science. 2015;**75**:1-14. DOI: 10.1016/j.paerosci.2015.02.001

[87] Krishnan KG, Milionis A, Loth E, Farrell TE, Crouch JD, Berry DH. Influence of hydrophobic and superhydrophobic surfaces on reducing aerodynamic insect residues. Applied Surface Science. 2017;**392**:723-731. DOI: 10.1016/j.apsusc.2016.09.096

[88] Kok M, Mertens T, Raps D, Young TM. Influence of surface characteristics on insect residue adhesion to aircraft leading edge surfaces. Progress in Organic Coatings. 2013;**76**:1567-1575. DOI: 10.1016/j.porgcoat.2013.06.013

Chapter 3

Superhydrophobic Membrane for Gas-Liquid Membrane Contactor Applications

Syafiqa M. Saleh, Athirah Mohd Tamidi, Farahdila Kadirkhan and Pei Ching Oh

Abstract

Membrane contactors allow for higher mass transfer per unit volume. Hence, there has been great interest in recent years on its development and applications in separation processes. It offers high interfacial area between liquid and gas phases while preventing direct mixture, and concurrently prevents flooding and foaming, thanks to the independent gas and liquid flow rates. However, wetting of the membrane pores is a serious problem for this technology application, where even partial membrane pores wetting could significantly deteriorate contactor performance. Therefore, it is crucial that the membranes are hydrophobic to maintain membrane nonwetting during operation. Moreover, any membrane surface modification to increase its hydrophobicity must also be highly stable, does not leach out, and can be applied for long-term operation. This chapter looks at the research done on superhydrophobic membranes for gas-liquid membrane contactor application and its recent advances.

Keywords: superhydrophobic, membrane, gas-liquid contactor, surface modification, wetting, polymeric

1. Introduction

There is a massive demand in recent years to develop new separation technologies or improve on existing ones, in order to enhance product quality, solve environmental issues, improve energy efficiency, increase process safety and/ or cost reduction. However, to compete with proven conventional separation technologies, critical attention must be given to every detail to achieve the required properties. In one such case, membrane technology has the potential to replace traditional energy-intensive separation techniques and provide a reliable option for sustainable industrial growth [1]. Today, membrane contactor has attained considerable attention due to its wide range of potential applications. The number of publications on the field has risen steadily since the past 20 years, as shown in **Figure 1** [2].

Since most of chemical separation processes are related to the contact of two different phases (liquid-liquid or gas-liquid), a membrane contactor system is deemed suitable for operations such as gas absorption and stripping, liquid-liquid extraction, distillation,

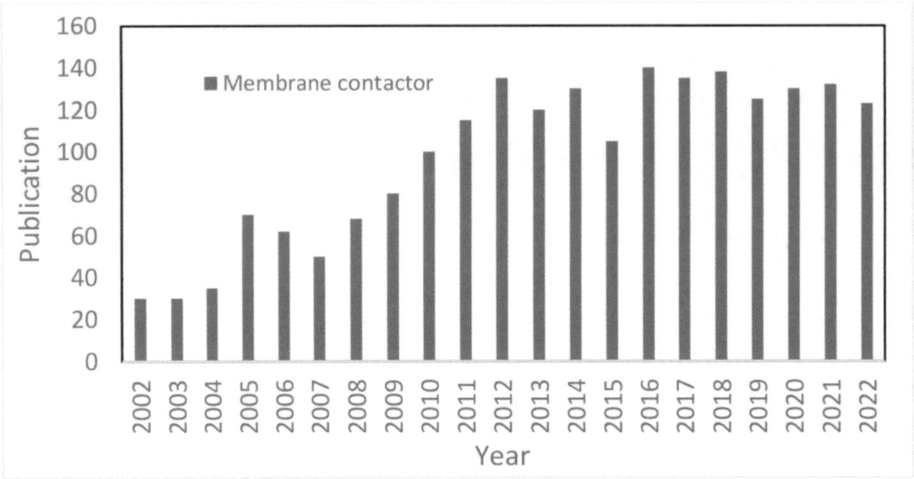

Figure 1.
The yearly publications on membrane contactor for the past 20 years shows immense interest in the field. Adopted from Ref. [2].

heterogeneous reactions, emulsification, demulsification, humidification, and dehumidification. There have been quite a number of commercialized membrane contactor system reported, such as ammonia recovery, de-gassing of water, and air humidification [3].

2. Membrane contactor: working mechanism and applications

2.1 Introduction to membrane contactor

A membrane contactor is a specialized device that can be used in various industrial processes and applications to transfer materials (such as gases or liquids) across a porous membrane between two phases. The porous membrane acts as a nonselective barrier between the phases, in which the transport of species through the membrane pores occurs by diffusion mechanism. In a membrane contactor system, the selectivity of the separation process is governed by the component's affinity to the specific phase. The detailed fundamentals and mass transfer transport mechanism of the species through porous membrane contactors can be found in the open literature [4, 5].

The primary purpose of a membrane contactor is to enhance mass transfer, separation, or purification processes. Membrane contactors offer several advantages over traditional contacting devices. They provide an exceptionally high interfacial contact area between the phases, which results in an improved mass transfer coefficient (K) and is a more compact separation system with less capital cost and energy consumption. This is contributed to the use of thousands of hollow membrane fibers in the module. For comparison, based on reported experimental results, the volume of a conventional packed column is 4–10 times bigger than a membrane contactor unit [6, 7].

Another advantage of membrane contactor systems is its operational flexibility. Since the membrane separates the two different phases, independent control of the phase's flow rates prevents operational problems like flooding, foaming, weeping, entrainment, and loading [4]. In addition, the modularity of membrane contactors leads to easy scale-up and scale-down. The characteristics of membrane contactors compared with conventional contactors are shown in **Table 1**.

Contactor type	Surface area/ volume (m^2/m^3)	Mass transfer coefficient ($K_l.a$) (s^{-1}) $\times 10^{-2}$	Gas/liquid volume flow (%)
Bubble column	50–600	0.5–12	60–98
Packed column	10–350	0.04–7	2–25
Venturi scrubber	150–2500	8–25	5–30
Membrane contactor	1000–10,000	5–50	1–99

Table 1.
Characteristics of different gas-liquid contactors (Ref. [8]).

Membrane contactors, however, do face some challenges. Regular asymmetric membranes have high gas permeability, ultra-thin skin layer, and small thickness; hence any resistance is negligible. In membrane contactor process on the other hand, the membrane phase itself adds extra resistance to mass transfer process. This resistance can be minimized using appropriate materials and optimizing fabrication conditions to improve the membrane structure and properties [8, 9].

Another crucial matter is that wetting of the membrane pores is a serious problem for this technology application. Partial wetting of the membrane can significantly deteriorate the mass transfer coefficient and decrease the separation efficiency [10, 11]. Several attempts have been made to develop and modify the membrane structure to minimize wetting and the mass transfer resistance; these are highlighted in the proceeding sections. In order to successfully implement membrane contactor technology in the industry, current status, and future direction of the technology are discussed critically in this chapter.

2.2 Applications of gas-liquid membrane contactor

The schematic of the mass transfer mechanism of the gas-liquid membrane contactor technology is shown in **Figure 2**. The overall mass transfer resistance, K is the combination of the gas (g) phase, membrane (m) and liquid (l) phase resistances that is shown in Eq. (1).

$$\frac{1}{K_o} = \frac{1}{K_m} + \frac{1}{K_l} + \frac{1}{K_g} \tag{1}$$

In an ideal case, the membrane pores are considered gas-filled, which minimizes the overall resistance. In fact, the nonwetted mode can be achieved by using highly hydrophobic or superhydrophobic membranes with small pore sizes. Partial wetting of the membrane can seriously increase the overall resistance and deteriorate the separation performance of the membrane [10]. The gas phase resistance can be reduced by increasing velocity and/or the species concentration. Using low-thickness porous hydrophobic membranes with ultra-thin skin layers may minimize the resistance for different applications. So far, several attempts have been reported in the literature to enhance the membrane structure for contactor applications [8, 9, 12]. In addition, introducing chemical reactions for separation purposes and increasing the liquid phase velocity can significantly enhance the liquid mass transfer coefficient and minimize the resistance [11, 13].

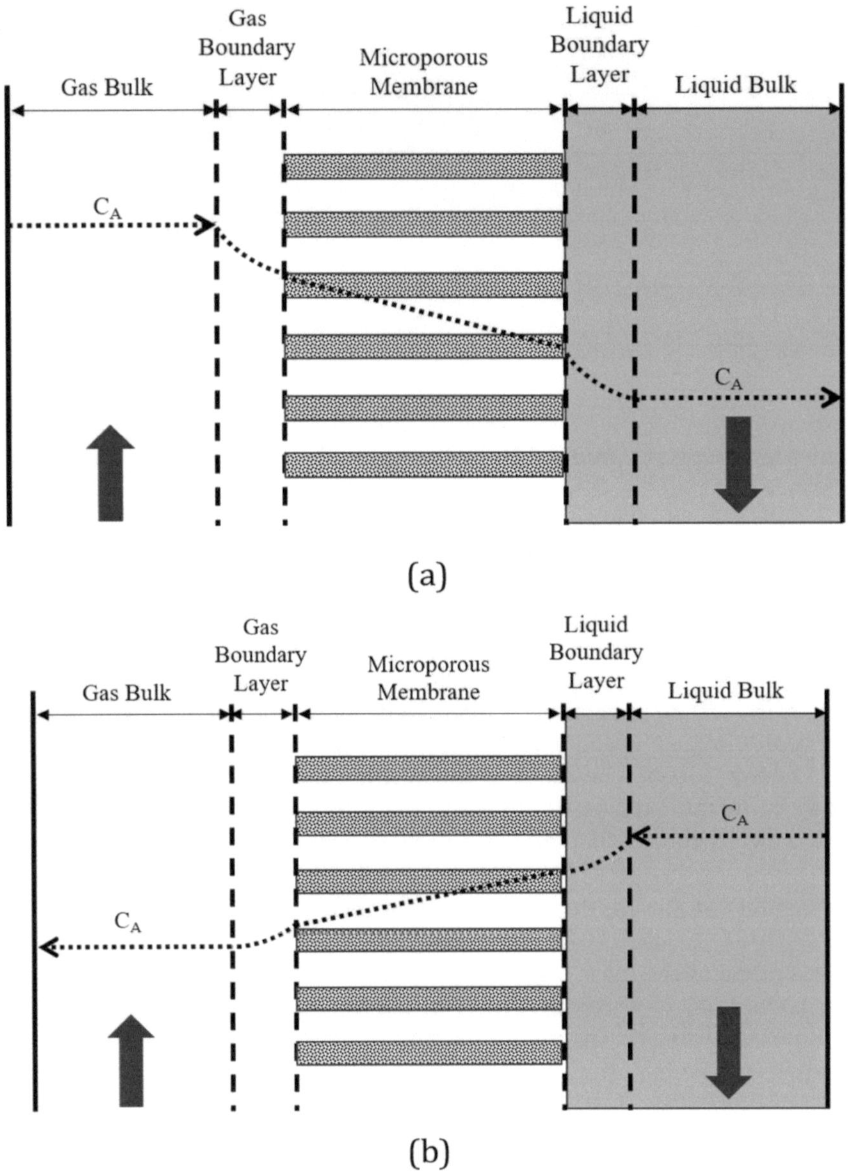

Figure 2.
Schematic of nonwetted mode mass transfer profile of component "A" (C_A) for different applications of gas-liquid membrane contactor: (a) gas absorption and dehumidification; and (b) stripping, degassing, humidification, distillation.

Since contact of two phases is possible via the membrane contactor, several applications can be expected in the scope of gas-liquid contacting processes. The use of gas-liquid membrane contactors covers a wide range of separation processes such as absorption/stripping, degassing, humidification/dehumidification and distillation. The varied applications and research on gas-liquid membrane contactors are summarized in **Table 2**.

As summarized in **Table 2**, so far, simultaneous absorption of various types of acid gases and toxic vapors such as VOCs, CO_2, H_2S, SO_2, and NO_2 with different liquid

Application	Gas	Liquid	Membrane	Research focus	References
Absorption	CO_2, SO_2, H_2S, NO_2, Ozone, VOCs, H_2S/ CO_2, CO_2/ SO_2, ethylene/ ethane	Water, Na_2SO_3, Na_2CO_3, $NaHCO_3$ and NaOH solutions, MEA, DEA, MDEA solutions and mixtures of amine, $AgNO_3$ solutions	PVDF, PSF and PEI, PP, PTFE, ePTFE, and composite PP.	• Development of hollow fiber membrane structure • Evaluation of different physical and chemical liquid absorbents for gas absorption • Simultaneous absorption of different gases • Process optimization • Process modeling and simulation	[14–18]
Stripping	CO_2/N_2, VOCs/ Air, ethylene/ ethane/ Air	Water, MEA and DEA solutions, silicon oil 200 fluid and Paratherm NF®, $AgNO_3$ solutions	PVDF, PSF, PEI, ceramic, PTFE and PP	• Development of membrane structure for improving gas stripping performance • Modeling of membrane contactor gas stripping • Effect of main operating parameters on stripping performance	[19–21]
Degassing	Air/O_2, N_2/O_2,	Water	PP, poly-4-methylpenten-1, silicone, rubber capillary (SILASTIC®),	• Applications for boiler feed water deoxygenation, ultrapure water production, and microbiocontamination. • Mass transfer analysis and process simulation	[22, 23]
Distillation	Water vapor, N_2	Aqueous mixtures of ethanol, isopropanol solution	Commercial PTFE (POREFLON), PP and PVDF	• Isopropanol separation in sweep gas membrane distillation and Ethanol separation in vacuum membrane distillation (VMD) were conducted. • Effect of operating conditions on flux improvement	[24–26]
Humidification	air	Water	Composite PVDF, In-house made PES and PSF,	• Model development of heat and mass transfer • The effects of structure-induced flow maldistributions on the deteriorations of humidification efficiencies. • Effect of operating parameters • Effect of membrane structure on humidification	[27, 28]
dehumidification	Air/ vapor	LiCl solution, Triethylene	Composite PVDF, Polydimethylsiloxane, (PDMS), and polyvinyltrimethylsilane	• Spiral wound and flat frame membrane modules • Mathematical modeling of heat and mass transfer in	[29, 30]

Application	Gas	Liquid	Membrane	Research focus	References
		glycol (TEG),	(PVTMS), PE, coated PEI	cross-flow hollow fiber membrane. • Parameters affecting vapor mass flux through the membrane. • Transversal flow modules	

Table 2.
Applications of gas/vapor-liquid membrane contactors.

absorbents have been conducted using membrane contactors [14–18]. Many researchers focused on CO_2 removal due to its greenhouse effect and impact to global warming. In correspondence, gas-stripping membrane contactors are developed to regenerate the liquid absorbents either as integrated process or standalone system [19–21]. Degassing to remove O_2 and CO_2 through membrane contactor modules from boiler feed water was developed to control corrosion [22, 23]. There is also vast interest in its application for membrane distillation of organic solvents [24–26]. Additionally, membrane contactor is also considered an alternative for air humidity control [27–30].

3. Current issues and challenges of membrane contactor

Membrane wetting is the main challenge of gas-liquid membrane contactors, as it can deteriorate the separation performance. Although the membrane employed in gas absorption is often hydrophobic and capable of resisting the wetting of absorbents, aqueous absorbent solutions of organic compounds, especially alkanolamines [31], could still penetrate the membrane pores over prolonged operation. The schematic diagram of nonwetting, overall-wetting, and partial wetting of the membrane contactor is shown in **Figure 3**.

The partial-wetting mode will cause the overall mass transfer resistance to increase rapidly and significantly influence the stability of long-term operation [10]. Malek et al. [33] investigated the effects of membrane-wetting pressures on the overall mass transfer. They concluded that partial wetting of the membrane can change the overall mass transfer coefficient of membrane gas absorption. Lu et al. [32] presented the wetting mechanism for hydrophobic hollow fiber membrane-alkanol amine absorbents for CO_2 capture. A mathematical model for wetting mechanism was proposed by correlating the resistance-in-series equation, the Laplace equation, and the pore size distribution function of the membrane.

For a given hydrophobic membrane material and structure, the degree of partial wetting of membrane pores depends on the surface tension the contact angle between the absorbents and the membrane surface, and the operating conditions. The minimum or breakthrough pressure (ΔP), exerted on the liquid phase to enter the membrane pore can be estimated by the Laplace-Young equation, as shown in Eq. (2) [34]:

$$\Delta P = \frac{2\gamma cos\theta}{r_{p,max}} \quad (2)$$

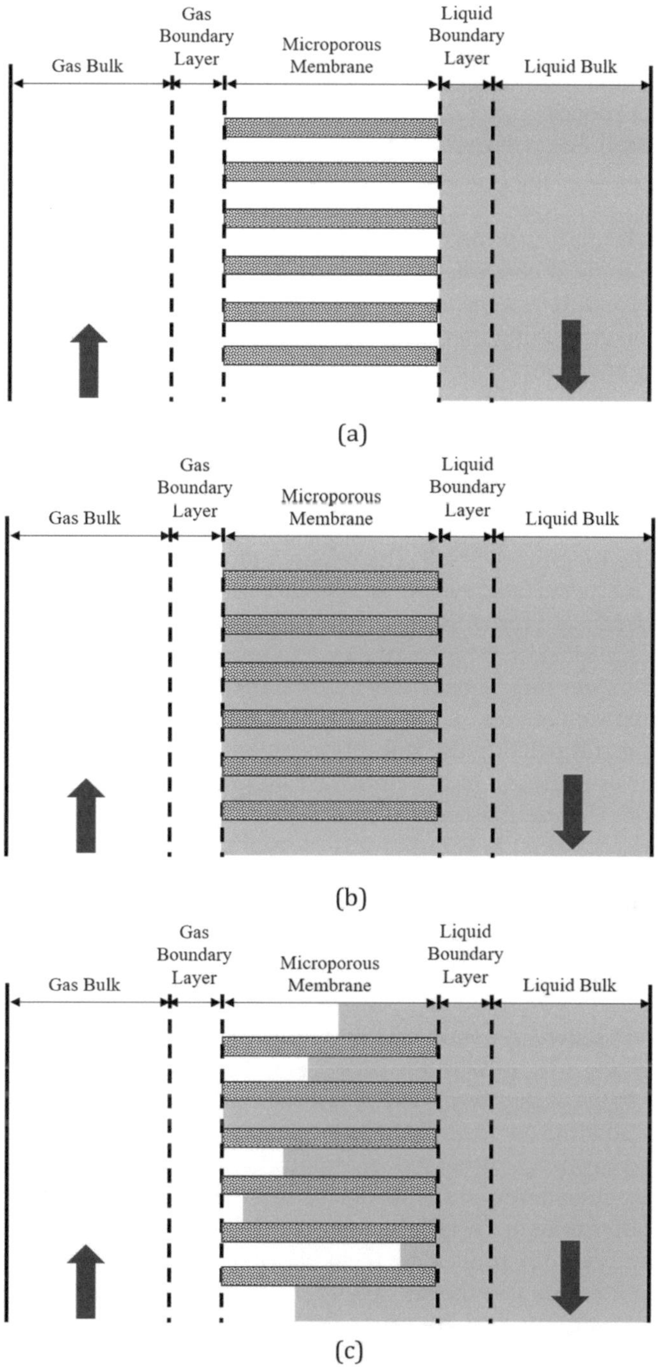

Figure 3.
Operation modes in a hydrophobic microporous hollow fiber membrane and pore-wetting patterns: (a) nonwetting mode, (b) overall-wetting mode, and (c) partial-wetting mode. Adapted from Ref. [32].

where γ is the surface tension of the liquid, θ the contact angle between the fluid phase and the membrane, and $r_{p,max}$ is the maximum membrane pore radius.

Thus, based on the Laplace-Young equation, membrane and liquid absorbent properties can influence membrane wetting [34]. Surface hydrophobicity and pore size are two essential membrane properties that can be controlled during fabrication and modification processes.

Among hydrophobic polymers, PP, PTFE, and PVDF have extensively been used for porous membrane fabrication [8, 13, 14]. Microporous PTFE membrane shows good gas absorption performance and long-term stability [35]; however, its application is limited due to its high production cost and unavailability in small diameters [12]. PP and PTFE are however not soluble in common solvents at room temperature and these membranes are normally fabricated using thermal and stretching methods. These methods do not have the flexibility to control pore size and porosity. PVDF, on the other hand, can be dissolved in organic solvents at room temperature, resulting in asymmetric membranes with controlled pore size, porosity, and low mass transfer resistance (favorable factors for membrane contactor) via phase inversion process. There have been extensive studies on improving membrane hydrophobicity, which will be explored further in detail in the next section.

Aside from the membrane itself, the selected liquid absorbent properties are also crucial, where its surface tension is key, affecting membrane wetting. Amine solutions have been used for acid gas capture extensively. These organic solutions possess low surface tension, which can easily penetrate through the membrane pores and result in wetting. Thus, there are studies to develop liquid with high reactivity and surface tension, like ionic liquids and amino acid salt solutions, to minimize wetting and prolong the stability of the operation [17, 36, 37]. The liquid absorbents are tailored to have high affinity to the gas component, high compatibility with the membrane material, as well as low vapor pressure and viscosity. Low vapor pressure is important to minimize liquid loss during the operation, while low viscosity is required to minimize the pressure drop in the membrane contactor module. These tailored properties would increase mass transfer and enhance the membrane contactor separation performance. New and novel absorbents, however, suffer from high cost and low regenerability for a cost-effective operation.

The interaction between membrane material and liquid absorbent can significantly affect the membrane structure and properties, which could result in unviable long-term operation. Other than inert PTFE, materials such as PP and PVDF have shown interaction with alkaline solutions [38, 39], which are usually used as liquid absorbents for acid gas capture. Therefore, compatibility of the absorbent with the membrane needs to be considered in the design of long-term operation.

Aside from polymeric material, the idea of using composite membranes has also been explored due to its long-term wetting stability and high compatibility with liquid absorbents [40–42]. Composite ceramic membranes have high thermal stability and are proposed for high-temperature processes like gas-stripping membrane contactors. Thin-coated layers are developed to minimize membrane resistance, attain high gas permeability, high hydrophobicity to minimize wetting, high mechanical resistance against erosion of the liquid flow, and high chemical resistance for a prolonged operation. They provide a good alternative for membrane contactor application besides polymeric membranes.

4. Membrane fabrication and modification for superhydrophobicity improvement

A surface is considered superhydrophobic when the static water contact angle exceeds 150° and has low contact angle hysteresis (i.e., less than 10°) [43]. A superhydrophobic surface can be created by reducing the surface free energy of a rough surface or roughening a low surface energy material or a combination of both [44, 45]. The higher the surface roughness, the solid-liquid contact surface would become smaller and hence larger contact angle [46].

One tactic to obtain superhydrophobic membrane is through fabricating a membrane with a rough surface directly, i.e., by having nanoparticles entrapped in the membrane structure by adding them into the polymer matrix itself [47]. The other approach is to modify fabricated membrane surface with low surface energy materials [48]. **Figure 4** shows a general summary of the fabrication methods of superhydrophobic polymeric membranes.

4.1 Membrane fabrication with nanoparticles to achieve superhydrophobicity

One way of achieving superhydrophobic properties is to have nanoparticles entrapped in the membrane structure by adding them into the polymer matrix itself [47]; this has been intensively studied to improve the membrane properties, such as permeability, selectivity, and physical strength [49]. 2 wt% of nanoparticles is reported to improve significantly the membrane contact angle by 18% [50]. The addition of nanoparticles with longer hydrophobic chains is considered to provide higher membrane hydrophobicity [50, 51]. Blending modification is the most practical way that can be applied to industrial-scale production [52].

Another way to achieve membranes with different structures is to adjust the bath formula and conditions during nonsolvent-induced phase inversion (NIPS). PVDF precipitation is controlled by both liquid-liquid demixing and solid-liquid demixing, which could result in amorphous cellular morphology or interlinked semi-crystalline

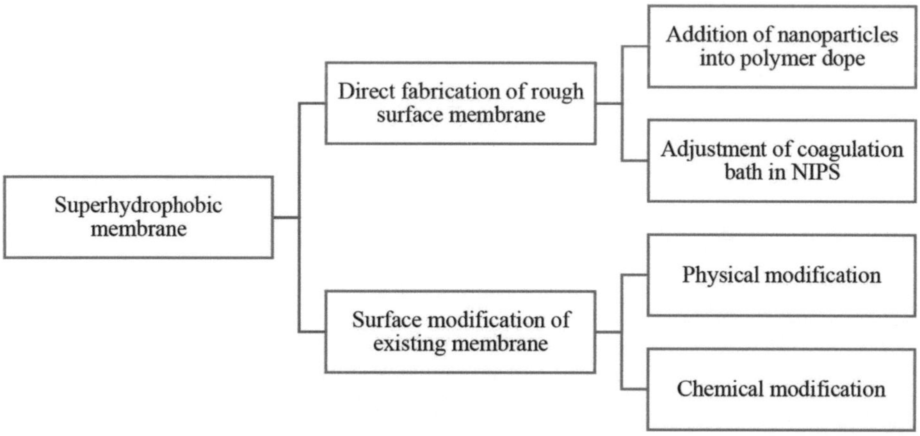

Figure 4.
Summary of the fabrication methods of superhydrophobic polymeric membranes.

Membrane material	Nanoparticles	Fabrication	Contact angle	Surface free energy (mN/m)	Surface roughness (nm)	Application	Reference
PVDF	Hydrophobic modified SiO_2 nanoparticles in ethanol bath	NIPS	164°	0.9	773	CO_2 absorption	[53]
PVDF	Silica-hexadecyltrimethoxysilane (SiO2-HDTMS) in dope	NIPS	160°	—	—	CO_2 absorption	[54]
PVDF	Crystallization of pristine PVDF (P-PVDF) powder and defluorinated PVDF (D-PVDF) powder	Solute and solvent co-crystallization	151°	—	451	Membrane distillation	[55]
PVDF-HFP	Nano-SiO_2 particles	NIPS	151°	3.49	3590	Gas-liquid separation	[56]
PVDF	Poly-dimethylsilane-grafted-silica (PGS) nanoparticles in coagulation bath	NIPS	161°	0.68	3860	Gas-liquid separation	[57]
PVDF	TiO_2-SiO_2 mixture	NIPS	161°	—	870	Membrane distillation	[58]
Polyvinylidene luoride-co-hexafluoropropylene (PcH)	Electrospin with carbon nanotubes (CNTs)	Electrospin	158°	—	455	Membrane distillation	[59]

Table 3.
Fabrication of superhydrophobic membrane.

particle structure, respectively. Hence, the coagulation bath significantly affects the characteristics of the fabricated PVDF membranes [53].

Table 3 shows some of the methods used by researchers to produce superhydrophobic membranes for gas-liquid contact applications. Researchers also reported a significant reduction in surface free energy and increased surface roughness.

The biggest issue is, however, the agglomeration that takes place when blending nanoparticles together with polymer. Agglomeration causes the instability of casting solution and nonuniform distribution of particles in the membranes; this leads to a change in membrane topography, microstructure, and performance [60]. Pang et al. managed to obtain evenly distribution of SiO_2 in their membrane by introducing hydroxyl groups into PVDF, which enhanced the affinity between the PVDF chains and the hydrophilic nano-SiO_2 particles [54].

4.2 Membrane surface modification for superhydrophobicity

Surface modification methods can be either through physical modification or chemical modification. In physical modification, the modifiers exist on membrane surface via physical interaction but not covalent bonding. Hence, the chemical composition of the membrane remains unchanged. Surface coating/ deposition is a simple yet effective method for membrane modification; however, interaction between the coated layer and the membrane surface is relatively weak, and probably lost during long-term operation [60].

A variety of methods, including the wet-chemical modification [61], plasma treatment [52, 62], sol-gel method [47, 63], and atomic layer deposition [64, 65], can be found in literature and have been used for the modification of membrane to improve its properties, as well as to functionalize the membrane with additional properties [61]. However, surface modification techniques using plasma treatments and particle beam irradiation need expensive and complex equipment, therefore they could not be used widely in industry [66].

For superhydrophobic coating of membranes, nanoparticles are commonly used, such as SiO_2, TiO_2, Al_2O_3, ZnO, and $CaCO_3$ [44], to roughen the surface. Polymers that possess hydroxyl groups or carboxyl groups on their molecular chains form bonding under certain conditions with the hydroxyl groups on the surface of inorganic nanoparticles like nano SiO_2 and nano TiO_2 [67]. Some researchers took the approach to lower surface energy to achieve superhydrophobicity; for example, Xiong et al. [68] used surface silanization and fluorination to lower the surface energy to only one-third of the original PTFE membrane, and Meng et al. [43] used low surface energy of fluorinated silanes with surface roughness of nanoparticles.

Table 4 describes some of the surface modification methods used to attain superhydrophobicity on polymeric membrane.

Membrane material	Modification method	Contact angle	Surface roughness (nm)	Membrane Application	Reference
PP	Coating with granular PP. and the mixtures of Methyl ethyl ketone (MEK) and cyclohexanone	158°	274	CO_2 absorption	[12]
PVDF	Coating with TiO2 and 1H,1H,2H,2H-perfluorododecyl-trichlorosilane (FTCS) with PEG template	160°	—	Membrane distillation	[43]
PVDF	Dip coat with TiO_2-FTCS	158°	1380	Membrane distillation	[48]
PTFE	Surface silanization and fluorination	159°	5380	Membrane distillation	[68]
PVDF	Catechol(CA)/polyethyleneimine (PEI) co-deposition, sol-gel growth of SiO_2 nanoparticles, and grafting of long chain perfluorosilane	157°	133	Membrane distillation	[69]
PP	UV grafting and immobilsation of octadecyltrichlorosilane (OTS)	162°	126	Water-in-oil emulsions separation	[70]
PTFE	Spray-deposition with mixture of sillca and MEK	158°	210	CO_2 absorption	[71]
PVDF	Thiol-ene click chemistry reaction with silica and PDMS	157°	191	Dehumidification	[72]
PP	Coating with modified SiO2 nanoparticles	163°	4538	Membrane distillation	[73]
Polyurethane acrylate (PUA)	Spray coat with functionalized SiO_2	150°	—	Oil/Water Separation	[74]

Table 4.
Surface modification of polymeric membrane achieving superhydrophobicity.

4.3 Current separation performance of superhydrophobic membrane

Performance of superhydrophobic membrane for contactor application is compared in this section. For CO_2 removal process, the normally reported range of CO_2 flux for polymeric membrane contactors is 0.2–6.0 mmol/m^2.s [3]. It is worth noting that CO_2 removal using membrane contactor performance is much more complex and influenced by a lot of factors, like the choice of absorbent and operating conditions (i.e., flowrate, temperature, and pressure) [75]. However, for ease of comparison, flux is a valid criterion to monitor and evaluate its performance since it is used as a scale-up factor for the technology [35]. **Table 5** shows the reported performance of the produced superhydrophobic membrane (listed in column "modified") and compares it against the flux for membrane in pristine condition.

As observed from **Table 5**, the produced superhydrophobic membrane in general provides better separation performance. The results above describe the performance during short testing, and the effect of superhydrophobicity is actually more substantial for long-term operation. For example, the modified PP/cyclohexanone:MEK membrane, during the initial operating period it, performs poorer than pristine membrane because although the modification treatment enhanced hydrophobicity, it also decreased membrane porosity and increased its thickness. However, after 20 days of operation, modified PP/cyclohexanone:MEK membrane CO_2 flux is stable at 1.3 mmol/m^2.s; the pristine membrane on the other hand declined by 78% down to 0.5 mmol/ m^2.s [12].

Another considerable application of membrane contactor is for membrane distillation. As shown in **Table 6**, the difference can be seen in the membrane distillation performance between pristine membranes and modified superhydrophobic membranes.

Similar to membrane contactors for CO_2 removal, the significance of superhydrophobic membranes for membrane distillation application is much more impactful after long-duration testing. PVDF-TiO_2-FTCS-PEG membrane remained stable at 12 kg/m^2.h after 24 hours, while the pristine membrane flux dropped by 44% to 10 kg/m^2.h [43]. Therefore, superhydrophobicity is a good strategy to minimize wetting and improve membrane contactors for industrial applications.

4.4 Stability of superhydrophobic membrane

Although superhydrophobic membrane is less prone to wetting, chemical reaction between the surface of the membrane and the absorbent could still occur

Superhydrophobic membrane	Gas flow rate (ml/min)	Solution	Solution flow rate (ml/min)	CO_2 Flux (mmol/ m^2.s)		Reference
				Pristine	Modified	
PP/cyclohexanone: MEK	200 (20% CO_2)	MEA	17	2.3	1.8	[12]
HMSNs/PVDF	1500 (pureCO_2)	12 wt% MEA	90	40	50	[53]
PVDF-SiO2-HDTMS	20 (19% CO_2)	DEA	50	1.1	2.4	[54]
PTFE-silica MEK	1000	K_2CO_3	75	1.5	1.85	[71]

Table 5.
CO_2 removal performance of superhydrophobic membrane in contactor process.

Superhydrophobic membrane	Feed flow rate (ml/min)	Feed composition	Flux ($kg/m^2.h$)		Reference
			Pristine	Modified	
PVDF-TiO_2-FTCS-PEG	300	10 wt% NaCl	18	14	[43]
PVDF ENM-TiO_2-FTCS	750	3.5 wt% NaCl	46.2	73.4	[48]
P-PVDF/D-PVDF	—	35 g/L NaCl	—	68.43	[55]
PVDF/TiO_2-SiO_2	300	0.1 mmol/L gallic acid solution with surfactant	10	15	[58]
CNT/PVDF-co-HFP	400	30 g/L NaCl solution	28.8	29.5	[59]
F-SiO_2@PTFE	150	5.5 wt% NaCl	7	9	[68]
F-SiO_2/PEIC/PVDF	900	35 g/L NaCl	14.12	14.58	[69]
OTS-PP	1667	35 g/L NaCl	5.0	8.5	[73]

Table 6.
Membrane distillation performance of superhydrophobic membrane.

after prolonged contact, leading to decline in membrane stability and decrease in its mass transfer flux [54]. Hence, it is important to evaluate its performance for long duration. For superhydrophobic membrane as contactor, the decline in performance could be mainly due to chemical degradation of the interface [54] or damage to the coating [73].

Long-term testing of these superhydrophobic membranes for gas-liquid contactor for more than 24 hours is scarcely found in literature. **Table 7** lists the reported long-duration performances that are done at least for a day. Despite a slight decline, the results are encouraging and pave the way for potential developments.

Aside from the decline in performance, the membrane hydrophobicity is also observed to have reduced, for example PVDF-SiO_2-HDTMS contact angle reduced from 160° to 154° after 20 days of operation. Chemical degradation of the interface by MEA solvent leads to the dehydrogenation of the PVDF surface. This is more destructive for pristine PVDF where the contact angle suffers greatly and dropped to 59° [51].

5. Challenges and way forward

This chapter has thus far described the methods of superhydrophobic polymeric membrane fabrication, which are either direct fabrication of rough surface membrane, or surface modification of existing membrane. Drawing insights from pertinent literature, these methods demonstrate the capability to achieve an impressive contact angle ranging from 150° to 164° (as described in Section 4.1 and 4.2). Promising membranes for application should have outstanding antiwettability not only on outside surfaces but internal surfaces as well, i.e., the surface of pores and channels, for persistence of the antiwettability in service [53]. For effective scale-up for industrial applications, there would be a trade-off between the best superhydrophobic membrane and the most practical method of membrane fabrication and modification.

Even though research has been conducted to fabricate superhydrophobic membranes successfully, there are still other challenges prior to further scale-up and

Membrane	Application	Long-duration performance	Reference
PP-MEK	CO_2 absorption with MEA	Decline 14% after 20 days	[12]
PVDF-SiO_2-HDTMS	CO_2 absorption with DEA	Decline 3% after 20 days	[54]
PVDF + LDPE	CO_2 absorption with MEA	Decline 14% after 1 day	[54]
Modified PTFE	Membrane distillation	Decline 11% after 1 day	[68]
OTS-PP	Membrane distillation	Decline 13% after 3.5 days	[73]

Table 7.
Long-duration stability testing of superhydrophobic membrane for gas-liquid contactor application.

technology commercialization. As described by Lv et al. [12], the focus on increasing hydrophobicity is good for performance, but this alone would not offset the disadvantages of low porosity and high resistance. Hence, several strategies have been employed to reduce mass transfer resistance. One of the strategies is to braid, weave, or loom together the hollow fiber membranes to reduce liquid boundary layer resistance and improve mass transfer.

To overcome chemical degradation, the right selection and compatibility of the membrane material and solvent are crucial. This chapter has mainly focused on polymeric material; however, ceramic membranes are also possible alternative for membrane contactor application [40].

It is understandable that there is limitation, and only so far that improvement from superhydrophobic membrane can help. Thus, it is crucial for membrane contactor research to also optimize operating parameters to enhance performance. Surface tension of organic solvent decreases at higher temperature, leading to membrane

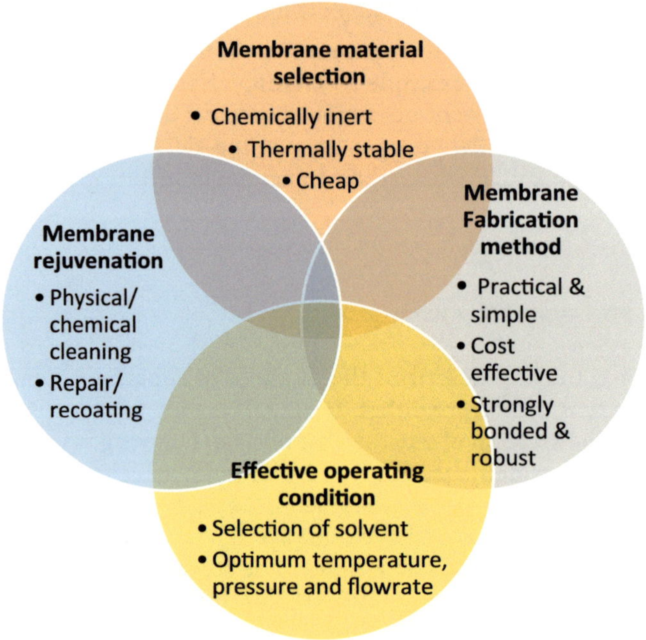

Figure 5.
Summary of challenges and strategies on superhydrophobic membrane for gas-liquid membrane contactor applications.

wetting; therefore, operating at lower temperature would be an option. Good trans-membrane pressure control during operation is very important to maintain a slight higher pressure at the liquid side, but not high enough to wet the membrane. In addition, it is essential to conduct extended stability testing lasting beyond a few days to confidently move these researches from proof-of-concept to prototype demonstration at the actual site.

Another potential strategy is to consider methods of rejuvenating the used membrane. Restoring the functionality of the membrane is a good alternative that could extend the lifespan and reduce the cost of operation (i.e., instead of replacing it with new membranes). Physical cleaning may be done by simply washing with water and blowing with gas to clear the clogged pores. Other chemical cleaning or repairs could be considered depending on the membrane and application.

The summary of challenges and future directions regarding superhydrophobic membranes for gas-liquid membrane contactor applications are highlighted in **Figure 5**.

Acknowledgements

The authors would like to gratefully acknowledge the support from PETRONAS Research Sdn. Bhd. and Universiti Teknology PETRONAS.

Conflict of interest

The authors declare no conflict of interest.

Author details

Syafiqa M. Saleh[1]*, Athirah Mohd Tamidi[1], Farahdila Kadirkhan[1] and Pei Ching Oh[2]

1 Carbon Capture, Utilization & Storage (CCUS) R&D Department, PETRONAS Research Sdn. Bhd, Jalan Ayer Itam, Kawasan Institusi Bangi, Kajang, Malaysia

2 CO_2 Research Centre (CO_2RES), Chemical Engineering Department, Universiti Teknologi PETRONAS, Bandar Seri Iskandar, Malaysia

*Address all correspondence to: syafiqa_saleh@petronas.com.my

References

[1] Drioli E, Curcio E, Di Profio G. State of the art and recent progresses in membrane contactors. Chemical Engineering Research and Design. 2005; **83**(3A):223-233. DOI: 10.1205/cherd.04203

[2] Lee Y, Park YJ, Lee J, Bae TH. Recent advances and emerging applications of membrane contactors. Chemical Engineering Journal. 2023;**461** (November 2022):141948. DOI: 10.1016/j.cej.2023.141948

[3] Kim S, Scholes CA, Heath DE, Kentish SE. Gas-liquid membrane contactors for carbon dioxide separation : A review. Chemical Engineering Journal. 2021;**411**(January): 128468. DOI: 10.1016/j.cej.2021.128468

[4] Mansourizadeh A, Ismail AF. Hollow fiber gas – Liquid membrane contactors for acid gas capture : A review. Journal of Hazardous Materials. 2009;**171**:38-53. DOI: 10.1016/j.jhazmat.2009.06.026

[5] Li J-L, Chen B-H. Review of CO_2 absorption using chemical solvents in hollow fiber membrane contactors. Separation and Purification Technology. 2005;**41**(2):109-122. DOI: 10.1016/j.seppur.2004.09.008

[6] deMontigny D, Tontiwachwuthikul P, Chakma A. Comparing the absorption performance of packed columns and membrane contactors. Industrial and Engineering Chemistry Research. 2005; **44**:5726-5732

[7] Yeon SH, Sea B, Park YI, Lee KH. Determination of mass transfer rates in PVDF and PTFE hollow fiber membranes for CO_2 absorption. Separation Science and Technology. 2003;**38**:271-293

[8] Mansourizadeh A, Fauzi A. A developed asymmetric PVDF hollow fiber membrane structure for CO_2 absorption. International Journal of Greenhouse Gas Control. 2011;**5**(2): 374-380

[9] Gugliluzza A, Drioli E. PVDF and HYFLON AD membranes: Ideal interfaces for contactor applications. Journal of Membrane Science. 2007;**300**: 51-62

[10] Wang R, Zhang HY, Feron PHM, Liang DT. Influence of membrane wetting on CO_2 capture in microporous hollow fiber membrane contactors. Separation and Purification Technology. 2005;**46**(1–2):33-40. DOI: 10.1016/j.seppur.2005.04.007

[11] Keshavarz P, Fathikalajahi J, Ayatollahi S. Analysis of CO_2 separation and simulation of a partially wetted hollow fiber membrane contactor. Journal of Hazardous Materials. 2008; **152**:1237-1247

[12] Lv Y, Yu X, Jia J, Tu ST, Yan J, Dahlquist E. Fabrication and characterization of superhydrophobic polypropylene hollow fiber membranes for carbon dioxide absorption. Applied Energy. 2012;**90**(1):167-174. DOI: 10.1016/j.apenergy.2010.12.038

[13] deMontigny D, Tontiwachwuthikul P, Chakma A. Using polypropylene and polytetrafluoroethylene membranes in a membrane contactor for CO_2 absorption. Journal of Membrane Science. 2006;**277** (1–2):99-107. DOI: 10.1016/j.memsci.2005.10.024

[14] Mansourizadeh A, Aslmahdavi Z, Ismail AF, Matsuura T. Blend polyvinylidene fluoride/surface modifying macromolecule hollow fiber

membrane contactors for CO_2 absorption. International Journal of Greenhouse Gas Control. 2014;**26**: 83-92

[15] Tamidi AM, Yasin NH. CO_2 and H_2S co-removal from natural gas using membrane contactor technology. In: SPE Asia Pacific Oil and Gas Conference and Exhibition. Virtual: SPE; 2020. DOI: 10.2118/202269-ms

[16] Tamidi AM, Saleh SM, Chian QV, Hanafiah R. Integrated absorption and regeneration membrane contactor technology for removal of CO_2 and H_2S from natural gas. In: Proceedings of Abu Dhabi International Petroleum Exhibition & Conference (ADIPEC 2022); October 31–November 3 2022. Abu Dhabi, UAE: OnePetro; 2022. DOI: 10.2118/211216-MS

[17] Luis P, Garea A, Irabien A. Zero solvent emission process for sulfur dioxide recovery using a membrane contactor and ionic liquids. Journal of Membrane Science. 2009;**330**(1–2): 80-89

[18] Sun X, Meng F, Yang F. Application of seawater to enhance SO2 removal from simulated flue gas through hollow fiber membrane contacto. Journal of Membrane Science. 2007;**312** (1–2):6-14

[19] Chan ZP, Li L, Kang G, Ab Manan N, Cao Y, Wang T. Regeneration of CO_2 absorbent with membrane contactor via pressure swing. Chemical Engineering Research and Design. 2021; **167**:107-115. DOI: 10.1016/j.cherd.2021.01.007

[20] Poddar TK, Majumdar S, Sirkar KK. Removal of VOCs from air by membrane-based absorption and stripping. Journal of Membrane Science. 1996;**120**(2):221-237

[21] Rahbari-Sisakht M, Ismail AF, Rana D, Matsuura T. Carbon dioxide stripping from diethanolamine solution through porous surface modified PVDF hollow fiber membrane contactor. Journal of Membrane Science. 2013;**427**: 270-275

[22] Peng ZG, Lee SH, Zhou T, Shieh JJ, Chung TS. A study on pilot-scale degassing by polypropylene (PP) hollow fiber membrane contactors. Desalination. 2008;**234**(1–3):316-322

[23] Tan X, Capar G, Li K. Analysis of dissolved oxygen removal in hollow fibre membrane modules: Effect of water vapour. Journal of Membrane Science. 2004;**251**(1–2):111-119

[24] Izquierdo-Gil MA, Jonsson G. Factors affecting flux and ethanol separation performance in vacuum membrane distillation (VMD). Journal of Membrane Science. 2003;**214**(1):113-130

[25] Lee CH, Hong WH. Effect of operating variables on the flux and selectivity in sweep gas membrane distillation for dilute aqueous isopropanol. Journal of Membrane Science. 2001;**188**(1):79-86

[26] EL-Bourawi EL, Khayet M, Godino MP. Application of vacuum membrane distillation for ammonia removal. Journal of Membrane Science. 2007;**301**(1–2):200-209

[27] Zhang LZ, Huang SM. Coupled heat and mass transfer in a counter flow hollow fiber membrane module for air humidification. International Journal of Heat and Mass Transfer. 2011;**54**(5–6): 1055-1063

[28] Bergero S, Chiari A. Experimental and theoretical analysis of air humidification/dehumidification processes using hydrophobic capillary

contactors. Applied Thermal Engineering. 2001;**21**(11):1119-1135

[29] Isetti C, Nannei E, Magrini A. On the application of a membrane air-liquid contactor for air dehumidification. Energy and Buildings. 1997;**25**(3): 185-193

[30] Kneiful K, Nowak S, Albrecht W, Hilke R, Just R, Peinemann KV. Hollow fiber membrane contactor for air humidity control: Modules and membranes. Journal of Membrane Science. 2006;**276**(1–2):241-251

[31] Kumar PS, Hogendoorn JA, Feron PHM, Versteeg GF. New absorption liquids for the removal of CO 2 from dilute gas streams using membrane contactors. Chemical Engineering Science. 2002;**57**(9): 1639-1651

[32] Lu JG, Zheng YF, Cheng MD. Wetting mechanism in mass transfer process of hydrophobic membrane gas absorption. Journal of Membrane Science. 2008;**308**(1–2):180-190. DOI: 10.1016/j.memsci.2007.09.051

[33] Malek A, Li K, Teo WK. Modeling of microporous hollow fiber membrane modules operated under partially wetted conditions. Industrial and Engineering Chemistry Research. 1997;**36**(3):784-793

[34] Franken ACM, Nolten JAM, Mulder MHV, Bargeman D, Smolders CA. Wetting criteria for the applicability of membrane distillation. Journal of Membrane Science. 1987; **33**(3):315-328

[35] Chian V, Shah N, Chachuat B. Modeling for design and operation of high-pressure membrane contactors in natural gas sweetening. Chemical Engineering Research and Design. 2018; **132**:1005-1019. DOI: 10.1016/j.cherd.2018.01.033

[36] Feron PHM, Jansen AE. CO_2 separation with polyolefin membrane contactors and dedicated absorption liquids: Performances and prospects. Separation and Purification Technology. 2002;**27**(3):231-242

[37] Lu J, Zheng Y, Cheng M. Membrane contactor for CO_2 absorption applying amino-acid salt solutions. Desalination. 2009;**249**(2):498-502

[38] Wang R, Li DF, Zhou C, Liu M, Liang DT. Impact of DEA solutions with and without CO_2 loading on porous polypropylene membranes intended for use as contactors. Journal of Membrane Science. 2004;**229**(1–2):147-157

[39] Lv Y, Yu X, Tu S-T, Yan J, Dahlquist E. Wetting of polypropylene hollow fiber membrane contactors. Journal of Membrane Science. 2010;**362** (1–2):444-452. DOI: 10.1016/j.memsci.2010.06.067

[40] Koonaphapdeelert S, Li K. The development of ceramic hollow fibre membranes for a membrane contactor. Desalination. 2006;**200**(1–3):581-583

[41] Koonaphapdeelert S, Li K. Preparation and characterization of hydrophobic ceramic hollow fibre membrane. Journal of Materials Science. 2006;**291**(1–2):70-76

[42] Koonaphapdeelert S, Wu Z, Li K. Carbon dioxide stripping in ceramic hollow fibre membrane contactors. Chemical Engineering Science. 2008; **64**(1):1-8

[43] Meng S, Mansouri J, Ye Y, Chen V. Effect of templating agents on the properties and membrane distillation performance of TiO 2-coated PVDF

membranes. Journal of Membrane Science. 2014;**450**:48-59. DOI: 10.1016/j.memsci.2013.08.036

[44] Syafiq A, Pandey AK, Balakrishnan V, Rahim NA. Study on self-cleaning performance and hydrophobicity of TiO_2/silane coatings. Pigment and Resin Technology. 2018. DOI: 10.1108/PRT-02-2018-0010

[45] Razmjou A, Arifin E, Dong G, Mansouri J, Chen V. Superhydrophobic modification of TiO 2 nanocomposite PVDF membranes for applications in membrane distillation. Journal of Membrane Science. 2012;**415–416**: 850-863. DOI: 10.1016/j.memsci.2012.06.004

[46] Cao H et al. ZIF-8 based dual scale superhydrophobic membrane for membrane distillation. Desalination. 2023;**550**(January):116373. DOI: 10.1016/j.desal.2023.116373

[47] Zakria HS, Othman MHD, Kamaludin R, Kadir SHSA, Kurniawan TA, Jilani A. Immobilization techniques of a photocatalyst into and onto a polymer membrane for photocatalytic activity. RSC Advances. 2021;**11**:6985-7014. DOI: 10.1039/d0ra10964a

[48] Ren L, Xia F, Chen V, Shao J, Chen R, He Y. TiO 2—FTCS modified superhydrophobic PVDF electrospun nano fi brous membrane for desalination by direct contact membrane distillation. Desalination. 2017;**423**(August):1-11. DOI: 10.1016/j.desal.2017.09.004

[49] Zhang X, Wang DK, Diniz JC. Recent progresses on fabrication of photocatalytic membranes for water treatment. Catalysis Today. 2014;**230**: 47-54. DOI: 10.1016/j.cattod.2013.11.019

[50] Toh MJ, Oh PC, Chew TL, Ahmad AL. Preparation of polydimethylsiloxane-SiO2/PVDF-HFP mixed matrix membrane of enhanced wetting resistance for membrane gas absorption. Separation and Purification Technology. 2020;**244**(November 2019):116543. DOI: 10.1016/j.seppur.2020.116543

[51] Purcar V, Cinteza O, Ghiurea M, Balan A, Caprarescu S, Donescu D. Influence of hydrophobic characteristic of organo-modified precursor on wettability of silica film. Bulletin of Materials Science. 2014;**37**(1):107-115. DOI: 10.1007/s12034-014-0628-7

[52] Liu F, Hashim NA, Liu Y, Abed MRM, Li K. Progress in the production and modification of PVDF membranes. Journal of Membrane Science. 2011;**375**:1-27. DOI: 10.1016/j.memsci.2011.03.014

[53] Wu X et al. Superhydrophobic PVDF membrane induced by hydrophobic SiO2 nanoparticles and its use for CO_2 absorption. Separation and Purification Technology. 2018;**190**(July 2017): 108-116. DOI: 10.1016/j.seppur.2017.07.076

[54] Pang H, Qiu Y, Sheng W. Long-term stability of PVDF-SiO2-HDTMS composite hollow fiber membrane for carbon dioxide absorption in gas–liquid contacting process. Scientific Reports. 2023;**13**(1):1-10. DOI: 10.1038/s41598-023-31428-8

[55] Hou C, Du L, Li Y, Guo M, Zhou J, Qiao S. Superhydrophobic PVDF membrane formed by crystallization process for direct contact membrane distillation. iScience. 2023;**26**(5):106464. DOI: 10.1016/j.isci.2023.106464

[56] Toh MJ, Oh PC, Chew TL, Ahmad AL. Antiwettability enhancement of PVDF-HFP membrane via superhydrophobic modification by

SiO2 nanoparticles. Comptes Rendus Chimie. 2019;**22**(5):369-372. DOI: 10.1016/j.crci.2019.05.004

[57] Toh MJ, OH PC, Ahmad AAL, Caille J. Enhancing membrane wetting resistance through superhydrophobic modification by polydimethylsilane-grafted-SiO_2 nanoparticles. Korean Journal of Chemical Engineering. 2019; **36**(11):1854-1858. DOI: 10.1007/s11814-019-0362-3

[58] Hamzah N, Leo CP, Ooi BS. Superhydrophobic PVDF/TiO 2-SiO 2 membrane with hierarchical roughness in membrane distillation for water recovery from phenolic rich solution containing surfactant. Chinese Journal of Polymer Science. 2019;**37**:609-616. DOI: 10.1007/s10118-019-2235-y

[59] Tijing LD et al. Superhydrophobic nanofiber membrane containing carbon nanotubes for high-performance direct contact membrane distillation. Journal of Membrane Science. 2016;**502**:158-170. DOI: 10.1016/j.memsci.2015.12.014

[60] Kang G, Cao Y. Application and modification of PVDF membranes – A review. Journal of Membrane Science. 2014;**463**:145-165. DOI: 10.1016/j.memsci.2014.03.055

[61] Feng S, Zhong Z, Wang Y, Xing W, Drioli E. Progress and perspectives in PTFE membrane: Preparation, modification, and applications. Journal of Membrane Science. 2018;**549** (October 2017):332-349. DOI: 10.1016/j.memsci.2017.12.032

[62] Mathieson I. Pretreatments of Fluoropolymers to Enhance Adhesion [thesis]. Loughborough University; 1995

[63] Ola O, Maroto-valer MM. Review of material design and reactor engineering on TiO 2 photocatalysis for COu2 reduction. Journal of Photochemistry and Photobiology C: Photochemistry Reviews. 2015;**24**:16-42. DOI: 10.1016/j.jphotochemrev.2015.06.001

[64] Li N et al. Precisely-controlled modi fi cation of PVDF membranes with 3D TiO 2 / ZnO nanolayer : Enhanced anti-fouling performance by changing hydrophilicity and photocatalysis under visible light irradiation. Journal of Membrane Science. 2017;**528**(73): 359-368. DOI: 10.1016/j.memsci.2017.01.048

[65] Yang H-C, Waldman R, Chen Z, Darling S. Atomic layer deposition for membrane interface engineering. Nanoscale. 2018;**10**:20505-20513. DOI: 10.1039/C8NR08114J

[66] Wang S, Li J, Suo J, Luo T. Surface modification of porous poly (tetrafluoraethylene) film by a simple chemical oxidation treatment. Applied Surface Science. 2010;**256**:2293-2298. DOI: 10.1016/j.apsusc.2009.10.055

[67] Lei P et al. Immobilization of TiO 2 nanoparticles in polymeric substrates by chemical bonding for multi-cycle photodegradation of organic pollutants. Journal of Hazardous Materials. 2012; **227–228**:185-194. DOI: 10.1016/j.jhazmat.2012.05.029

[68] Xiong Z et al. Silanization enabled superhydrophobic PTFE membrane with antiwetting and antifouling properties for robust membrane distillation. Journal of Membrane Science. 2023;**674**(March): 121546. DOI: 10.1016/j.memsci.2023.121546

[69] Jin Y, Wei Z, Meng X, Li J. Fabrication of superhydrophobic PVDF membrane via catechol/polyamine co-deposition and SiO2 nanoparticles for membrane distillation. Journal of Environmental Chemical Engineering.

2023;**11**(3):110250. DOI: 10.1016/j.jece.2023.110250

[70] Wang Y et al. Superhydrophobic wrinkled skin grown on polypropylene membranes enhances oil-water emulsions separation. Journal of Environmental Chemical Engineering. 2023;**11**(3):110247. DOI: 10.1016/j.jece.2023.110247

[71] Li Y, Wang L, Hu X, Jin P, Song X. Surface modification to produce superhydrophobic hollow fiber membrane contactor to avoid membrane wetting for biogas purification under pressurized conditions. Separation and Purification Technology. 2018;**194** (November 2017):222-230. DOI: 10.1016/j.seppur.2017.11.041

[72] Jia H, Ge Y, Chen H, Zhang W, Lu X, Jia Y. Fabrication of superhydrophobic PVDF membrane based on thiol-ene click chemistry reaction for humidification. Journal of Materials Science. 2022;**57**:9202-9217

[73] Chang G, Liu Z, Guo X, Xiao C. Construction of superhydrophobic micro/nano-structure surface on polypropylene hollow fiber membrane and its application in membrane distillation. Journal of Materials Science. 2022;**57**(9):5658-5678. DOI: 10.1007/s10853-022-06942-1

[74] Kim DS, Kang J, Jung JY, Hwang M, Seo S, Kim JH. Facile fabrication of Superhydrophobic polymer membranes with hierarchical structure for efficient oil/water separation. Fibers and Polymers. 2022;**23**(9):2365-2372. DOI: 10.1007/s12221-022-0062-1

[75] Saleh SM, Razali A, Hanafiah R, Tamidi A, Chan ZP. Optimization of MDEA-PZ ratio and concentration for CO_2 removal in semi-lean membrane contactor process. In: Proceedings of International Conference on Process Engineering & Advance Materials (ICPEAM2020); 14-16 July 2017; Malaysia. E3S Web of Conferences 287, 03009. 2021. DOI:10.1051/e3sconf/202128703009

Chapter 4

The Effect of Micro/Nano Roughness on Antifouling and Bactericidal Surfaces

Reyna I. Garcia-Gonzalez and Colin R. Crick

Abstract

The importance of microorganisms, especially bacteria, has often been underestimated, yet they have vital roles in staying in the environment and affecting human health and industries. These microorganisms have complex systems and change quickly over time, becoming more resistant. The spread of harmful microorganisms has negative effects on industries and human health. Even microorganisms that seem harmless can be a big problem because they are becoming more resistant to normal cleaning and antibiotics. They resist ways like creating strong biofilms, which make these microorganisms even tougher and help infections spread. Although there are other options like using heat or chemicals, the problem of bacterial resistance is still a big worry for health and industries. Trying out new ideas that do not use chemicals or antibiotics, like using superhydrophobic surfaces, could be a big solution. These surfaces use both special chemicals and changes in how they feel to water to stop bacteria from sticking and growing. By looking for new ways, we can get better at dealing with these microorganisms and find better ways to live with them.

Keywords: roughness, bacteria, antimicrobial surfaces, self-cleaning surfaces, nanostructure, microstructure

1. Introduction

Single-celled organisms are often judged in everyday life to be insignificant or harmless. However, bacteria have been on Earth for billions of years and have been integral to generating and maintaining its habitability for plants and animals. Their complex metabolic systems and rapid evolutionary pathways help them become more and more resistant over time [1]. The complexity of single-celled organisms has been progressively revealed over decades of scientific research; without this, we would not have discovered their abilities, including their processes of developing resilience in a range of environments [2, 3]. The proliferation of harmful microorganisms affects a range of fields, including industrial processes that are affected by biofouling and corrosion. Microbes that present an intrinsically low risk may cause detrimental conditions when they are not attended to properly (e.g., thrush caused by yeast infections). Despite the widespread use of microbial elimination from surfaces, such as disinfectants and antibiotics, their proliferation is still problematic, as microbes have the capacity to develop resistance to these efforts. These resistance mechanisms make microbes more resilient to the local environment, generally through the generation of biofilms [4, 5]. Genetic and assembly modifications (i.e., biofilms) contribute to growing antibiotic and chemical resistance, which has presented a new challenge for scientists. Antimicrobial resistance is mainly driven by their inherent metabolic mechanisms: the diversity of genetics to create mutations, the capacity to exchange genetic information with other bacteria, biofilms formation, rapid reproduction, and forced exposure to antibiotics or chemicals that force genetic mutations or resistance genes [1, 6, 7]. To combat this issue, prudent and responsible use of antibiotics and chemicals as well as research and development of new antimicrobial agents are essential to slow down the development and spread of resistance. By consistently managing drugs and antibiotics, microbial resistance would be reduced worldwide, since the presence of these emerging contaminants helps the distribution of microbial-resistant agents. Over the years, there have been alternatives that avoid using chemical disinfectants and antibiotics, such as heat, ultraviolet light, ozone, colloidal silver, supercritical water treatment, and natural and phytochemical competitors [3, 8]. However, it is important to note that while these methods may be helpful in certain contexts, serious medical situations, or highly contaminated environments, antibiotics, and chemical disinfectants remain essential tools to fight infection and prevent the spread of pathogenic bacteria [2]. Antimicrobial resistance is a key reason for exploring biocide-free alternatives such as superhydrophobic surfaces, which focus on the removal of bacteria and microorganisms from the surfaces by preventing the attachment of organic material and preventing the accumulation of water, which could otherwise promote the growth and colonization of these microorganisms [9]. Superhydrophobic surfaces are made by altering their structural and chemical properties. An inherently water-repelling surface chemistry is required alongside a highly rough morphology (on both the micro and nano scales). Micro/nano roughness can be developed by forming bumps, pillars, ridges, or other types of surface structures. Additionally, a hierarchical arrangement can be created by incorporating multiple levels of micro and nanostructures.

2. Biofilm formation and bacterial attachment in roughness areas

Bacterial biofilm formation represents a dynamic phenomenon that allows bacterial cells to adhere to surfaces and aggregate into colonies. Through this process of biofilm formation, pathogenic bacteria can prosper by adjusting to their external environment [10–12]. This resilience is obtained by microbial biopolymers such as poly-gamma-glutamic acid, proteins, exopolysaccharides, and extracellular DNA, which allows a more suitable environment to be created [11, 13]. Pathogenic bacteria can grow as biofilms, protecting them against antimicrobial agents and other substances such as disinfectants and antibiotics. The combination of these factors allows bacteria to develop resistance to antibiotics and chemicals, posing significant challenges in healthcare, agriculture, and environmental protection [2, 10]. Therefore, avoiding the formation of these biofilms by developing superhydrophobic surfaces by micro and nanostructures is a pathway to research, and then bacteria resistance can be reduced. The agglomeration process of bacteria within a biofilm depends on several distinct stages where the implementation of a superhydrophobic material could affect their formation, avoiding microorganisms during the reversible adhesion stage [2, 4]. The main objective of this stage is to obstruct the continuation of biofilm formation because bacteria are not already attached to the surface; in this way, a superhydrophobic surface could limit bacteria to starting the second phase of the biofilm formation, the irreversible adhesion [13–16]. During the irreversible adhesion stage, the bacteria undergo a transformation characterized by the loss of their flagellum, rendering them immobile on the surface. Subsequently, these immobilized bacteria begin to secrete material, enabling them to adhere firmly to the surface and remain devoid of movement, creating connections and marking the initiation of the third stage in the biofilm process. The colonization stage requires a proliferation of bacteria in such a manner that, in conjunction with the excreted slime, they form a suitable structure to acquire external nutrients through water channels. This colonization process is vital for the bacteria's growth, stability, and development. It is enhanced by quorum sensing, which is the complex system of communication between bacteria to coordinate the behavior into the biofilm formation as well as regulate the gene expression to make it favorable to biofilm formation [4, 17]. Upon reaching a considerable size and nourishment level, the biofilm enters the active dispersal stage. During this phase, bacteria start to depart from the biofilm, equipped with flagella, to colonize new surfaces that have not yet been populated. This repetitive cycle (**Figure 1**) remains the biofilm formation process, which could be limited by superhydrophobic properties [15, 16].

As previously mentioned, the limited control of bacteria spread could lead to infections in various sectors, including healthcare, agriculture, and industrial processes. For instance, hospitals are a critical area of concern where nosocomial infections can result in prolonged stays, disabilities, and financial burdens [18]. In agriculture, the misuse of antibiotics in farm animals has contributed to bacteria resistance, such as Penicillin resistance in humans, which can be transmitted to humans through food consumption and contact [2, 19]. Regarding the industry field, biofilms can decrease heat transfer across surfaces, make fluids harder to flow, and even increase surface degradation. Moreover, biofilm formation of resistant microorganisms that grow in process reactors and pipelines can be spread and generate polluted food even after it has been processed, triggering infections in

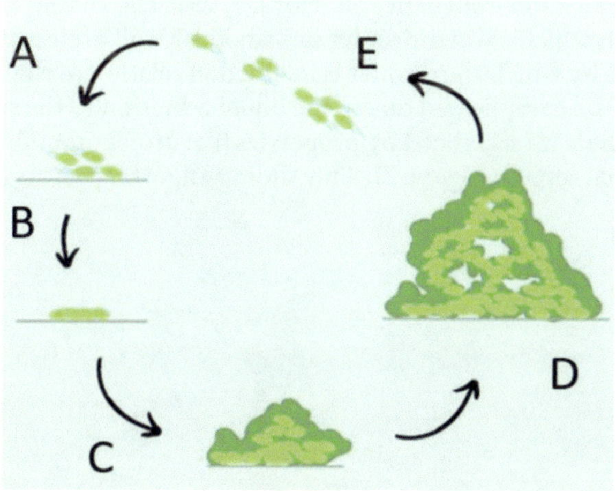

Figure 1.
Biofilm formation. (A) Reversible adhesion stage, (B) irreversible adhesion, (C) biofilm formation, (D) colonization process by quorum sensing, and (E) dispersal stage.

the purchasers [20, 21]. While bacterial contamination can indeed be mitigated through the implementation of singular or multi-tiered functionalization steps across surfaces to achieve remarkable antiseptic potential, preferably during the initial phases of bacterial adhesion preceding biofilm formation, an environmental alternative without the use of antibiotics and chemicals involves the development of superhydrophobic surfaces through structural modification, as it could be seen in nature [5, 12].

3. Roughness effect in hysteresis, tilt angle, and wettability in materials

A superhydrophobic surface is characterized by an extreme water repellency, namely the ability to repel water droplets to a remarkable extent, causing them to bead up and easily roll off the surface. Incorporating micro and nano-scale irregularities serves a dual purpose: enhancing water repellency (avoid adhesion) and amplifying the hysteresis property [22, 23]. The effect of low surface energy materials and the surface roughness trigger droplets to bead up into spherical droplets as they roll off the surface, carrying away dust and surface contaminants [24, 25]. The achieved roughness on the material improves the property of capturing air between the water droplet and the surface, minimizing the contact area and reducing the adhesion between water and the surface [23, 26]. As a result, water droplets cannot wet the surface effectively, leading to the water-repellent behavior observed on superhydrophobic surfaces. This property has many applications, including self-cleaning coatings, anti-icing surfaces, and even potential medical and industrial uses [23, 27]. Wettability varies depending on the surface's physical structure and chemical properties [28]. Measurement of surface wettability is principally achieved through the Water Contact Angle (WCA), sliding angle, and contact angle hysteresis (CAH).

The WCA describes the angle formed between the plane of the surface and the surface of the water droplet at the three-phase contact point (water-vapor-solid) [29, 30]. Varying these surface properties can result in tremendous changes to its water repellency. Essentially, it illustrates the capacity to which a surface repels or attracts a liquid. If the angle is large, the droplet tends to bead up and roll off the surface, indicating a super hydrophobic or water-repellent behavior [25, 29]. The sliding angle characterizes the facility with which a water droplet can smoothly roll off a surface, in this way, quantifies the angle by which the droplet is inclined in relation to the surface plane until movement starts to be complicated due to the liquid adhesion to the surface (32,37). Then, the sliding angle is influenced by properties like droplet mobility, adhesion, and resistance to external forces (**Figure 2**). This sliding angle reaches its highest point when

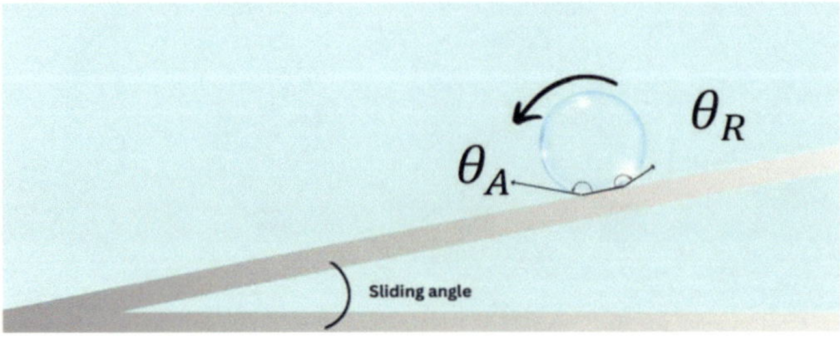

Figure 2.
Sliding angle and hysteresis. Water droplets roll off a tilted surface.

the droplet begins to roll off the surface, being the point at which the lowest receding contact angle (upper edge) and the highest advancing contact angle (lower edge) are reached [31].

$$CAH = (\theta_A) - (\theta_R) \tag{1}$$

The difference between the advancing angle (θ_A) and the receding angle (θ_R) is commonly referred to as the contact angle hysteresis (Eq. (1)) [32]. This hysteresis phenomenon in water contact angle measurements emerges due to the complex interactions between a liquid droplet and a solid surface during interactions that involve vapor, liquid, and solid surface tensions [33]. Surface roughness introduces irregularities that influence the behavior of the liquid droplet [24, 34]. Then, when having a highly hydrophobic material, the θ_A and θ_R get high values, and the difference between them is about zero.

The CAH phenomenon as well as the analysis of tilt angle and WCA are vital in real-world scenarios, influencing the behavior of liquids on surfaces during dynamic conditions [22, 33] and finding applications in fields ranging from self-cleaning materials to efficient fluid transport systems. Additionally, understanding and manipulating WCA, tilt angle, and CAH contributes to adapting surface functionalities to specific needs, developing innovations in numerous industries, such as the aerospace sector, where improved aerodynamics and anti-icing properties are sought, as well as the medical field, where avoiding wettability and biofilm formation on medical devices is crucial [27, 35, 36].

As these three parameters describe the relation between the surface properties and the liquid behavior, how specific surfaces repel water and are self-cleaned could be analyzed by Young's equation, Wenzel State, and the Cassie-Baxter State. The pioneering investigation into measuring a characteristic that determines whether a material is hydrophobic or not was initiated by Young, whose equation describes the behavior of a droplet when in contact with a surface; the droplet shape changes when it interacts on a surface, creating a new interface-three-phase contact line in an ideal smooth surface (**Figure 3**) [31, 37, 38]. Then, these three interfacial tensions (solid–liquid, liquid–gas, and solid–gas interfacial tensions) must be balanced to the total surface free energy to describe the corresponding contact angle [39]. However, to explain the behavior of a droplet by the WCA on a rough surface, the protuberances of these surfaces must be considered as the main feature [34, 40]. The roughness effect on the surfaces could be described by the fundamental states known as Wenzel State and Cassie-Baxter State to describe how a droplet of water interacts with a surface (**Figure 3**). Regarding the Wenzel state, when a droplet of water comes into contact with a surface, the liquid fills up the gaps present on the surface to completely cover them, leading to a complete contact with water surface [41–43]. This completely wetting surface could amplify the hydrophobic or hydrophilic properties of the surface, depending on whether the droplet creates a convex or concave shape. On the other hand, the Cassie–Baxter state suggests the opposite scenario; the liquid droplets do not fill these gaps created by the micro and nano roughness but rather by air, resulting in a surface composed of both solid and air [43–45]. As the droplet is suspended by air and some spots of the material, the liquid can easily slide off the surface due to low adhesion, preventing water and nutrient adhesion and thus decreasing bacterial growth. Through the interplay between these states, antibacterial and bactericidal properties could be achieved on different material surfaces.

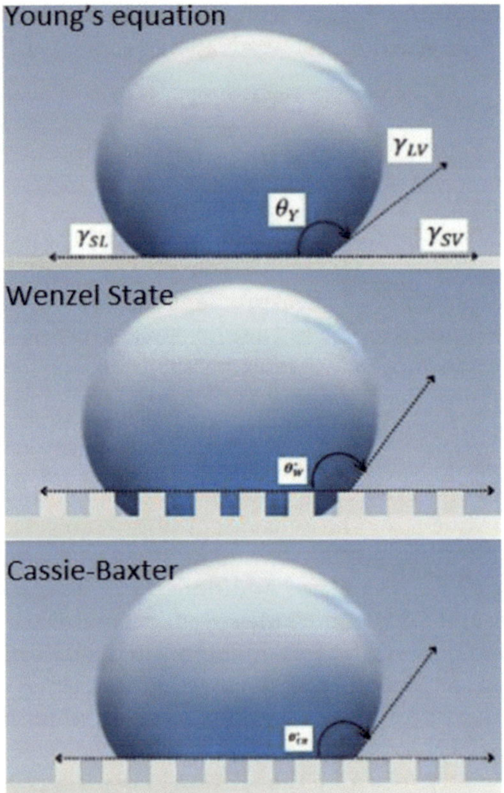

Figure 3.
Models that describe wettability.

4. Antibiofouling and bactericide properties

As described above, the hierarchical structure on surfaces increases the hydrophobicity of a material. However, bactericide and anti-biofouling properties greatly influence a material when industrial and environmental processes are being analyzed to prevent the spread of infections, corrosion, and biofouling. Even when both properties involve inhibiting or eliminating bacteria, they operate in distinct ways. Antibiofouling surfaces prevent the proliferation of bacteria, decreasing the availability of initial attachment of bacteria, while bactericidal surfaces directly kill bacteria [46]. The anti-biofouling property refers to a surface that can impede the growth and reproduction of bacteria by affecting the bacterial metabolic processes, disrupting cell wall formation, or even perturbing essential functions of certain cells [47–49]. Hence, the antibacterial principle is to avoid bacteria proliferating and decreasing biofilm formation by eliminating essential nutrient sources to the correct bacteria growth or adding elements that could affect the bacteria biofilm formation [47, 50]. The main nutrients required for bacteria to grow are water, carbon, nitrogen, and common minerals such as salts, phosphate, sulfate, magnesium, and calcium [51, 52]. The limitation of bacteria growth by changing the amount of nutrients was analyzed to study the impact of microorganisms on humans [53]. Pioneering work in creating a culture medium for the growth of specific bacteria was developed by Louis Pasteur in 1860 [51], which demonstrates the role of microorganisms in fermentation and diseases. However, Robert Koch developed the first solid culture to grow and isolate bacteria by a selection of nutrients required to make possible the individual study of metabolic processes of the different colonies of bacteria [54, 55]. This selective and differential media improves microbiology research by possibly differentiating microorganisms between them based on their metabolic requirements. The use of these different culture media has demonstrated that nutrient limitation could be a critical feature in slowing down bacteria growth on certain surfaces, making them valuable for disease prevention and hygiene improvements.

In the real world, nutrients could be limited in the environment, such as in the glacial environments, where bacteria face nutrient limitations due to the low organic matter content and slow nutrient release from ice melt [56]. Bacteria in desert soils are limited to rainfall and high evaporation rates; water becomes the main limitation for these bacteria [57]. These limitations of nucleotides required for DNA synthesis affect the DNA replication in the metabolisms of bacteria to the correct growth and cell division [58, 59]. The limitation of amino acids from organic matter restricts protein synthesis, which is one of the most important metabolic processes for enzyme activity, meaning essential cellular functions. In the same way, the cell wall formation can be compromised by the organic matter limitation, and the amino acids are the precursors of peptidoglycan synthesis that is primarily composed of bacterial cell walls [60, 61]. These are some examples of how nutrient limitations in removing organic matter and water from the surfaces could trigger more hygienic and clean surfaces through anti-biofouling properties. Hierarchical structures, with their inherent ability to repel water through hydrophobicity, offer a multifaceted solution that extends beyond surface water resistance. These structures possess a self-cleaning mechanism that extends to the removal of nutrients and dead cells of microorganisms that might have adhered to the surface. This self-cleaning capacity is a result of the interrelation between the surface's architecture and the water-repellent nature. The surfaces not only prevent water from adhering to them but also inhibit the attachment of organic matter, such as dead cells of microorganisms and nutrients, from remaining on the surface [62]. On the other hand, the bactericide property points out a more proactive approach to decrease bacteria growth and proliferation on surfaces, directly involving bacteria neutralization or killing bacteria on contact, which makes bacteria unable to form biofilms and incapable of reproduction [63, 64]. Even the bactericide surface is more effective when combining physical, chemical, and, in some cases, photonic mechanisms; the micro and nano roughness structures represent the main point of analysis for developing these bactericide structures [46, 62].

In their initial stages, designs for bactericidal surfaces predominantly relied on chemical methods as standalone approaches. These methods involved incorporating silver nanoparticles or applying antimicrobial compounds to the surface, effectively achieving their intended antibacterial effect [6, 49]. While these strategies have demonstrated success, they have also introduced certain challenges, like environmental concerns related to toxicity, as well as the gradual decrease in effectiveness over time due to the bacterial resistance and possible degradation and diminishing concentration of the active compounds [7, 65]. Furthermore, given the escalating concern about antimicrobial resistance, elaborating hierarchical structures is an alternative to research.

4.1 Antibacterial micro and nanostructures in the real world

The starting point of these structures are examples found in nature such as cicada wings, shark and springtail skin, and lotus leaves. Hierarchical structures made in laboratories and industry that effectively prevent bacteria growth find their inspiration in the remarkable designs found in nature. These natural micro and nano roughness surfaces, with unique advantages, serve as the foundation for innovative solutions in certain fields, including antibacterial technology [66–68]. Cicada wings exhibit a hierarchical arrangement characterized by micro- and nano-scale structures. These structures give the wings their water-repellent properties and prevent bacterial adhesion. The intricate patterns on cicada wings create a surface that is not only water-resistant but also impedes the agglomeration of organic matter and nutrients for bacteria growth [67, 69–71]. Bacteria can be eliminated easily from these cicada wings due to the combination of micro and nano protrusions that provide water-repelling characteristics, which affects their growth and colonization [46, 69, 72]. Similarly, shark skin poses a unique texture that reduces water friction. This texture consists of tiny square structures called dermal denticles, which serve as an inspiration for antibacterial surfaces due to hydrophobicity but also for antimicrobial properties [73–76]. In the same way, springtail skin features a pattern that enables it to regulate moisture levels and avoid immersion when standing on water [77, 78]. The lotus leaf, known for its exceptional self-cleaning properties due to micro and nanostructures, enhances water droplets to bead up and easily roll off, carrying away contaminants. This behavior, known as the *Lotus effect*, serves as a model for creating surfaces with both self-cleaning and antibacterial properties [36, 68, 78–80]. By incorporating these hierarchical structures into materials, the research seeks to not only prevent bacterial adhesion but also enable easy removal of any attached contaminants. Then, examples such as cicada wings, shark skin, springtail skin, and lotus leaves have exhibited the importance of complex patterns and textures in resisting bacterial colonization. By emulating these natural designs, researchers are developing innovative solutions for creating surfaces that are not only antibacterial but also self-cleaning, paving the way for cleaner and safer environments across various applications [70, 81]. These surfaces appear to effectively prevent biofilm formation and bacterial proliferation, offering a promising perspective to create cleaner and more hygienic materials in certain environments.

The comparison of structures found in nature can be compared, and the similarities suggest similar behavior through bacterial attachment. An example of these is the structure laboratory made with SiO_2 nanoparticles spread on the micropatterned wafers (Image A, B, C, D) and found in natural structures such as the lotus leaves (Image E and F) and springtail skin (Image G) in **Figure 4**. The micro and nano roughness of these surfaces achieves superhydrophobic properties as well as antibacterial and antifouling qualities.

5. Self-cleaning surfaces

The advantage that these superhydrophobic surfaces could achieve is that they not only could be cleaned easily but also, could have the property of being clean by themselves, self-cleaning properties allow droplets of water to carry out contaminants, mainly organic material, because of the low affinity to be adhered to the surface [83]. Moreover, by limiting the removal of nutrients from the surface, giving an anti-biofouling property, micro and nanostructures are able to affect bacterial cell walls and impede bacteria attachment. Adhesion of the metabolites improves the presence of biofilms that tend to create biofouling issues; therefore, their repulsion is highly important for long-term antifouling [84, 85]. Self-cleaning surfaces have

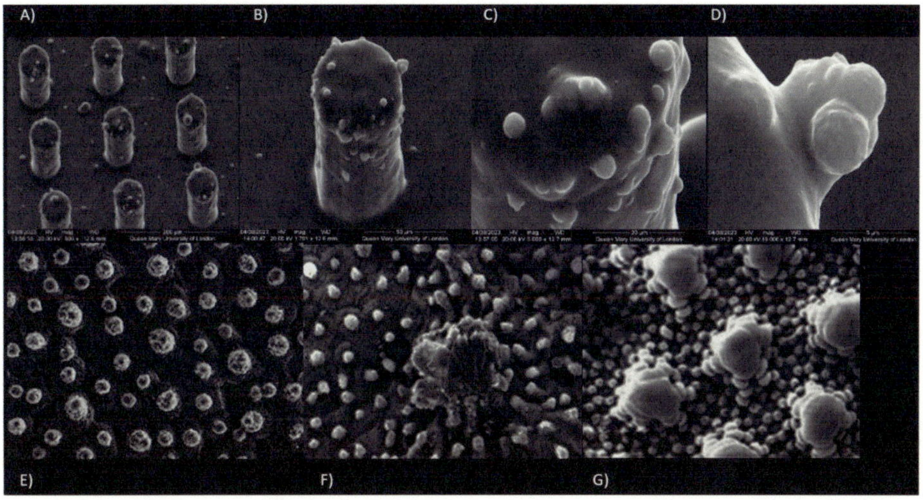

Figure 4.
Comparison of PDMS-covered pillars and SiO_2 nanoparticles and lotus leaf SEM (scanning electron microscopy) images. (A) Micropillars covered by SiO_2 nanoparticles, (B) a single micropillar covered by SiO_2 nanoparticles, (C) SiO_2 nanoparticles spread on the top of a micropillar, (D) SiO_2 nanoparticles cluster on the top of a micropillar, (E) lotus leaf, (F) lotus leave, (G) springtail skin. (lotus leaf image (E) reproduced from Zhao [63], lotus leaf image (F) and springtail skin image (G) reproduced from Yang et al. [82]).

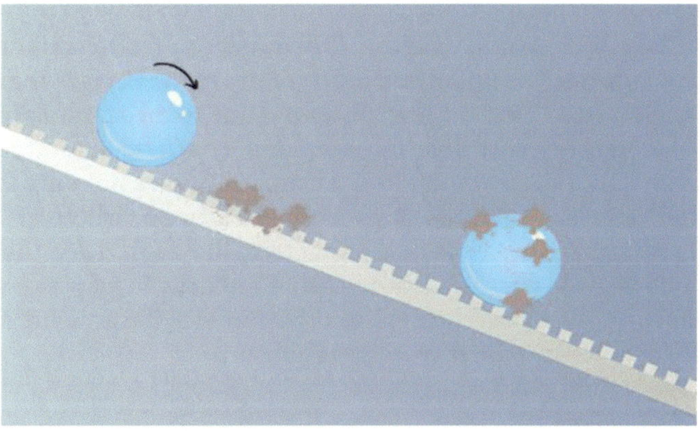

Figure 5.
Self-cleaning surfaces.

received a lot of attention in research as well as in commercial applications. Potential applications could be developed for self-cleaning surfaces such as automotive, building, household, optical applications, and aerospace [79, 84–87]. Therefore, to achieve an antibacterial effect, the surface's tilt angle must be small with a high contact angle hysteresis for droplets to fall out of the surface (**Figure 5**).

Moreover, by reaching a high hydrophobicity, bacteria will be attached to water by being more attracted by it instead of staying on the surface. Nevertheless, eliminating water from the surfaces will make it more difficult for bacteria to grow because of limited wettability [10, 11, 15, 21, 88]. As shown in Video 1 (https://drive.google.com/file/d/1IbIawbAZ1uAWTxSAPWQLF1rfcpBGea1E/view?usp=drive_link), when creating a self-cleaning surface by micro-patter (silicon pillars), the droplet of water

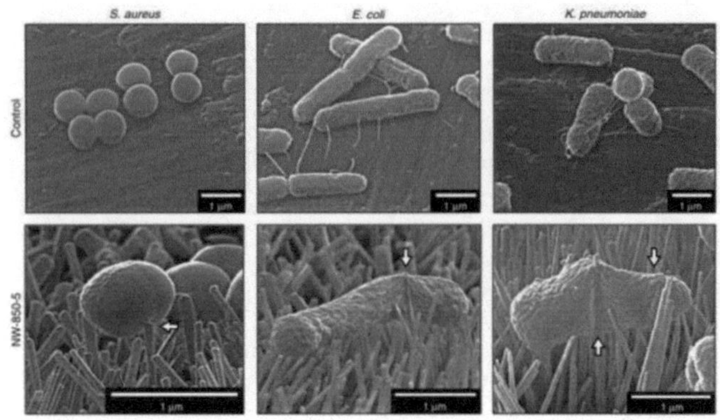

Figure 6.
SEM images highlight the interaction between TiO2 nanopillars and E. coli, K. Pneumonia, and S. aureus (reproduced from Jenkings et al. [63]).

appears to remove the dust particles (paper made) that remain on the top of a surface. Despite the water not being adhered to the surface material, the amount of water and the sliding angle required to roll off is noticeably more elevated compared to the results that a hierarchical structure could reach on the surface, as shown in Video 2 (https://drive.google.com/file/d/1eFLqx8dHFp_q9q4HMHXkhe5WHw3WTWrn/view?usp=drive_link). When combining micro-roughness (silicon pillars) and nano roughness on the top (SiO_2 nanoparticles), the property of self-cleaning surface is evident. This surface removes the undesired material from the surface remarkably with a lower sliding angle and an evident low hysteresis due to the stable spherical shape of the droplets of water. Another illustrative example of these laboratory-fabricated structures utilizing nanopillars resembling natural examples not only for antifouling properties but for bactericide purposes is shown in **Figure 6**, where TiO_2 nanopillars demonstrate to damage the cell walls of bacteria like *Escherichia coli* or *Klebsiella pneumoniae*, which leads to bacteria eradication. Similarly, *Staphylococcus aureus* poses a smaller morphology; the hierarchical structure impedes the attachment of coccus and disrupts the correct functionality of metabolism.

6. Bacteria on hydrophobic and hydrophilic surfaces

When comparing the behavior of bacteria on superhydrophilic and hydrophobic surfaces, their distinct wetting properties come into play. Superhydrophilic surfaces exhibit a strong affinity for water, causing water droplets to spread out extensively, creating a larger contact area. In the context of bacterial interaction, these surfaces encourage bacterial adherence and expansion, as the increased contact area enhances the chances of bacteria attaching to the surface [88]. This characteristic can be advantageous in scenarios such as wound healing and biofilm studies, where promoting bacterial colonization is desired. In contrast, hydrophobicity leads to water droplets forming compact beads that readily roll off. When bacteria encounter hydrophobic surfaces, their ability to adhere becomes compromised [89, 90]. The limited contact area and weak attractive forces inhibit bacterial attachment and colonization. This behavior is particularly useful in applications where preventing microbial contamination is crucial, such as medical devices and environments prone to high humidity. Hydrophobic surfaces effectively resist bacterial growth and biofilm formation by eliminating the main nutrients bacteria require to grow and avoiding water accumulation [89, 91].

Superhydrophilic surfaces facilitate bacterial adhesion and spreading, making them suitable for cases requiring enhanced bacterial colonization. On the other hand, hydrophobic surfaces deter bacterial attachment, proving advantageous when preventing bacterial growth and biofilm formation is the objective [63, 88–90].

7. Conclusion

The intricate process of bacterial biofilm formation presents challenges in various sectors due to its role in promoting bacterial resistance. The development of superhydrophobic surfaces, characterized by water repellence due to micro/nano-scale irregularities, inhibits biofilm formation and directly kills bacteria. This approach aligns with different stages of biofilm development, from avoiding initial adhesion to the surfaces before irreversible attachment to directly avoiding bacteria growth to reduce bacterial resistance. Superhydrophobic surfaces offer a potential solution by preventing water adherence and impeding biofilm formation, leading to improved hygiene and disease prevention in healthcare, agriculture, and industries. Furthermore, the development of hierarchical structures in surfaces offers solutions for anti-biofouling and bactericidal properties. These properties target bacterial attachment and metabolism, impede biofilm formation, and disrupt essential functions. The innovative combination of micro and nano roughness to reach superhydrophobicity and bacterial remotion presents a transformative approach to challenges in infection prevention, corrosion mitigation, and biofouling to achieve cleaner, safer, and more efficient environments. The exceptional potential of superhydrophobic surfaces developed by hierarchical structures provides beyond self-cleaning properties to even decrease bacteria growth and proliferation.

Acknowledgements

RIGG would like to acknowledge the Mexican-QMUL dual-center program for funding her master's research degree. CRC would also like to acknowledge the EPSRC for funding.

Conflict of interest

The authors have no conflicts to declare.

Author details

Reyna I. Garcia-Gonzalez* and Colin R. Crick
School of Engineering and Materials Science, Queen Mary University of London, London, United Kingdom

*Address all correspondence to: r.i.garcia-gonzalez@qmul.ac.uk

References

[1] Lawrence R, Jeyakumar E. Antimicrobial resistance: A cause for global concern. BMC Proceedings. 2013;7:S1. DOI: 10.1186/1753-6561-7-s3-s1

[2] Exner M, Bhattacharya S, Gebel J, Goroncy-Bermes P, Hartemann P, Heeg P, et al. Chemical disinfection in healthcare settings: Critical aspects for the development of global strategies. German Medical Science (GMS) hygiene and infection control, 15, Doc 36, German; 2020. DOI: 10.3205/dgkh000371

[3] Li W, Calle L, Hanford A, Stambaugh I, Callahan M. Investigation of silver biocide as a disinfection technology for spacecraft–an early literature review. In: 48th International Conference on Environmental Systems. Albuquerque, New Mexico. 2018. Available from: http://hdl.handle.net/2346/74083

[4] De Silva LADS, Heo GJ. Biofilm formation of pathogenic bacteria isolated from aquatic animals. Archives of Microbiology. 2023;**205**:36. DOI: 10.1007/s00203-022-03332-8

[5] Wang Y, Bian Z, Wang Y. Biofilm formation and inhibition mediated by bacterial quorum sensing. Applied Microbiology and Biotechnology. 2022;**106**:6365-6381. DOI: 10.1007/s00253-022-12150-3

[6] Mancuso G, Midiri A, Gerace E, Biondo C. Bacterial antibiotic resistance: The most critical pathogens. Pathogens. 2021;**10**:1310. DOI: 10.3390/pathogens10101310

[7] Coculescu B-I. Antimicrobial resistance induced by genetic changes. Journal of Medicine and Life. 2009;**2**(2):114-123

[8] Fong D, Gaulin C, Lê M-L, Shum M. Effectiveness of Alternative Antimicrobial Agents for Disinfection of Hard Surfaces. Public Health Agency of Canada; 2014

[9] Ashok D, Cheeseman S, Wang Y, Funnell B, Leung SF, Tricoli A, et al. Superhydrophobic surfaces to combat bacterial surface colonization. Advanced Materials Interfaces. 2023;**10**(24):1-22. DOI: 10.1002/admi.202300324

[10] Parrilli E, Tutino ML, Marino G. Biofilm as an adaptation strategy to extreme conditions. Rendiconti Lincei. 2022;**33**:527-536. DOI: 10.1007/s12210-022-01083-8

[11] Penesyan A, Paulsen IT, Kjelleberg S, Gillings MR. Three faces of biofilms: A microbial lifestyle, a nascent multicellular organism, and an incubator for diversity. npj Biofilms and Microbiomes. 2021;**7**:80. DOI: 10.1038/s41522-021-00251-2

[12] Hrynyshyn A, Simões M, Borges A. Biofilms in surgical site infections: Recent advances and novel prevention and eradication strategies. Antibiotics. 2022;**11**:69. DOI: 10.3390/antibiotics11010069

[13] Mohsin MZ, Omer R, Huang J, Mohsin A, Guo M, Qian J, et al. Advances in engineered *Bacillus subtilis* biofilms and spores, and their applications in bioremediation, biocatalysis, and biomaterials. Synthetic and Systems Biotechnology. 2021;**6**:180-191. DOI: 10.1016/j.synbio.2021.07.002

[14] Arnaouteli S, Bamford NC, Stanley-Wall NR, Kovács ÁT. *Bacillus subtilis* biofilm formation and social interactions. Nature Reviews

Microbiology. 2021;**19**:600-614. DOI: 10.1038/s41579-021-00540-9

[15] Bhattacharjee A, Khan M, Kleiman M, Hochbaum AI. Effects of growth surface topography on bacterial signaling in coculture biofilms. ACS Applied Materials & Interfaces. 2017;**9**:18531-18539. DOI: 10.1021/acsami.7b04223

[16] Sauer K, Stoodley P, Goeres DM, Hall-Stoodley L, Burmølle M, Stewart PS, et al. The biofilm life cycle: Expanding the conceptual model of biofilm formation. Nature Reviews. Microbiology. 2022;**20**:608-620. DOI: 10.1038/s41579-022-00767-0

[17] Haque S, Ahmad F, Dar SA, Jawed A, Mandal RK, Wahid M, et al. Developments in strategies for quorum sensing virulence factor inhibition to combat bacterial drug resistance. Microbial Pathogenesis. 2018;**121**:293-302. DOI: 10.1016/j.micpath.2018.05.046

[18] Khan HA, Baig FK, Mehboob R. Nosocomial infections: Epidemiology, prevention, control and surveillance. Asian Pacific Journal of Tropical Biomedicine. 2017;**7**:478-482. DOI: 10.1016/j.apjtb.2017.01.019

[19] Economou V, Gousia P. Agriculture and food animals as a source of antimicrobial-resistant bacteria. Infection and Drug Resistance. 2015;**8**:49-61. DOI: 10.2147/IDR.S55778

[20] Garrett TR, Bhakoo M, Zhang Z. Bacterial adhesion and biofilms on surfaces. Progress in Natural Science. 2008;**18**:1049-1056. DOI: 10.1016/j.pnsc.2008.04.001

[21] Vestby LK, Grønseth T, Simm R, Nesse LL. Bacterial biofilm and its role in the pathogenesis of disease. Antibiotics. 2020;**9**:59. DOI: 10.3390/antibiotics9020059

[22] Zhang Y, Zhang Z, Yang J, Yue Y, Zhang H. A review of recent advances in superhydrophobic surfaces and their applications in drag reduction and heat transfer. Nanomaterials. 2022;**12**:44. DOI: 10.3390/nano12010044

[23] Falde EJ, Yohe ST, Colson YL, Grinstaff MW. Superhydrophobic materials for biomedical applications. Biomaterials. 2016;**104**:87-103. DOI: 10.1016/j.biomaterials.2016.06.050

[24] Manoharan K, Bhattacharya S. Superhydrophobic surfaces review: Functional application, fabrication techniques and limitations. Journal of Micromanufacturing. 2019;**2**:59-78. DOI: 10.1177/2516598419836345

[25] Parvate S, Dixit P, Chattopadhyay S. Superhydrophobic surfaces: Insights from theory and experiment. Journal of Physical Chemistry B. 2020;**124**:1323-1360. DOI: 10.1021/acs.jpcb.9b08567

[26] Mohamed AMA, Abdullah AM, Younan NA. Corrosion behavior of superhydrophobic surfaces: A review. Arabian Journal of Chemistry. 2015;**8**:749-765. DOI: 10.1016/j.arabjc.2014.03.006

[27] Huang W, Huang J, Guo Z, Liu W. Icephobic/anti-icing properties of superhydrophobic surfaces. Advances in Colloid and Interface Science. 2022;**304**:102658. DOI: 10.1016/j.cis.2022.102658

[28] Slepickova, Kasalkova N, Slepicka P, Kolska Z, Svorcik V. Wettability and other surface properties of modified polymers. In: Wetting and Wettability. London, UK: IntechOpen; 2015. DOI: 10.5772/60824

[29] Celia E, Darmanin T, Taffin de Givenchy E, Amigoni S, Guittard F. Recent advances in designing superhydrophobic surfaces. Journal of Colloid and Interface Science. 2013;**402**:1-18. DOI: 10.1016/j.jcis.2013.03.041

[30] Genzer J, Efimenko K. Recent developments in superhydrophobic surfaces and their relevance to marine fouling: A review. Biofouling. 2006;**22**:339-360. DOI: 10.1080/08927010600980223

[31] Kim SH. Fabrication of superhydrophobic surfaces. Journal of Adhesion Science and Technology. 2008;**22**:235-250. DOI: 10.1163/156856108X305156

[32] Butt HJ, Liu J, Koynov K, Straub B, Hinduja C, Roismann I, et al. Contact angle hysteresis. Current Opinion in Colloid & Interface Science. 2022;**59**:101574. DOI: 10.1016/j.cocis.2022.101574

[33] Shams H, Basit K, Khan MA, Saleem S, Mansoor A. Realizing surface amphiphobicity using 3D printing techniques: A critical move towards manufacturing low-cost reentrant geometries. Additive Manufacturing. 2021;**38**:101777. DOI: 10.1016/j.addma.2020.101777

[34] Kanungo M, Mettu S, Law KY, Daniel S. Effect of roughness geometry on wetting and dewetting of rough PDMS surfaces. Langmuir. 2014;**30**:7358-7368. DOI: 10.1021/la404343n

[35] Gresham IJ, Neto C. Advances and challenges in slippery covalently-attached liquid surfaces. Advances in Colloid and Interface Science. 2023;**315**:102906. DOI: 10.1016/j.cis.2023.102906

[36] Avramescu RE, Ghica MV, Dinu-Pîrvu C, Prisada R, Popa L. Superhydrophobic natural and artificial surfaces-a structural approach. Materials. 2018;**11**:866. DOI: 10.3390/ma11050866

[37] Zhang L, Mehanna YA, Crick CR, Poole RJ. Surface tension and viscosity dependence of slip length over irregularly structured superhydrophobic surfaces. Langmuir. 2022;**38**:11873-11881. DOI: 10.1021/acs.langmuir.2c01323

[38] Calvimontes A. The measurement of the surface energy of solids using a laboratory drop tower. npj Microgravity. 2017;**3**:25. DOI: 10.1038/s41526-017-0031-y

[39] Law KY. Contact angle hysteresis on smooth/flat and rough surfaces. Interpretation, mechanism, and origin. Accounts of Materials Research. 2022;**3**:1-7. DOI: 10.1021/accountsmr.1c00051

[40] Mądry K, Nowicki W. Wetting between Cassie–Baxter and Wenzel regimes: A cellular model approach. European Physical Journal E. 2021;**44**:138. DOI: 10.1140/epje/s10189-021-00140-8

[41] Peters AM, Pirat C, Sbragaglia M, Borkent BM, Wessling M, Lohse D, et al. Cassie-Baxter to Wenzel state wetting transition: Scaling of the front velocity. European Physical Journal E. 2009;**29**:391-397. DOI: 10.1140/epje/i2009-10489-3

[42] Mohd Khairuddin FA, Rashid AA, Leo CP, Lim GK, Ahmad AL, Lim HM, et al. Recent progress in superhydrophobic rubber coatings. Progress in Organic Coating. 2022;**171**:107024. DOI: 10.1016/j.porgcoat.2022.107024

[43] Li J, Gong S, Zhang L, Cheng P, Ma X, Hong F. Wetting states and departure diameters of bubbles on micro-/nanostructured surfaces. Langmuir. 2022;**38**:3180-3188. DOI: 10.1021/acs.langmuir.1c03212

[44] Wang L, Tian Z, Jiang G, Luo X, Chen C, Hu X, et al. Spontaneous dewetting transitions of droplets during icing & melting cycle. Nature Communications. 2022;**13**:378. DOI: 10.1038/s41467-022-28036-x

[45] Catley TE, Corrigan RM, Parnell AJ. Designing effective antimicrobial nanostructured surfaces: Highlighting the lack of consensus in the literature. ACS Omega. 2023;**8**:14873-14883. DOI: 10.1021/acsomega.2c08068

[46] Razavi SMR, Oh J, Haasch RT, Kim K, Masoomi M, Bagheri R, et al. Environment-friendly antibiofouling superhydrophobic coatings. ACS Sustainable Chemistry & Engineering. 2019;**7**:14509-14520. DOI: 10.1021/acssuschemeng.9b02025

[47] He Z, Yang X, Wang N, Mu L, Pan J, Lan X, et al. Anti-biofouling polymers with special surface wettability for biomedical applications. Frontiers in Bioengineering and Biotechnology. 2021;**9**:807357. DOI: 10.3389/fbioe.2021.807357

[48] Imani SM, Maclachlan R, Rachwalski K, Chan Y, Lee B, McInnes M, et al. Flexible hierarchical wraps repel drug-resistant gram-negative and positive bacteria. ACS Nano. 2020;**14**:454-465. DOI: 10.1021/acsnano.9b06287

[49] Zhou H, Li Q, Zhang Z, Wang X, Niu H. Recent advances in superhydrophobic and antibacterial cellulose-based fibers and fabrics: Bioinspiration, strategies, and applications. Advanced Fiber Materials. Singapore: Springer Nature. 2023;**1**:1-37. DOI: 10.1007/s42765-023-00297-1

[50] Bonnet M, Lagier JC, Raoult D, Khelaifia S. Bacterial culture through selective and non-selective conditions: The evolution of culture media in clinical microbiology. New Microbes New Infections. 2020;**34**:100622. DOI: 10.1016/j.nmni.2019.100622

[51] Snell EE, Mitchell HK. Purine and pyrimidine bases as growth substances for lactic acid bacteria. Proceedings of the National Academy of Sciences. 1941;**27**(1):1-7

[52] Lagier JC, Hugon P, Khelaifia S, Fournier PE, La Scola B, Raoult D. The rebirth of culture in microbiology through the example of culturomics to study human gut microbiota. Clinical Microbiology Reviews. 2015;**28**:237-264. DOI: 10.1128/CMR.00014-14

[53] Zhang XH, Ahmad W, Zhu XY, Chen J, Austin B. Viable but nonculturable bacteria and their resuscitation: Implications for cultivating uncultured marine microorganisms. Marine Life Science & Technology. 2021;**3**:189-203. DOI: 10.1007/s42995-020-00041-3

[54] Lagier JC, Edouard S, Pagnier I, Mediannikov O, Drancourt M, Raoult D. Current and past strategies for bacterial culture in clinical microbiology. Clinical Microbiology Reviews. 2015;**28**:208-236. DOI: 10.1128/CMR.00110-14

[55] Larose C. Observatoire des Sciences de l'Univers de Grenoble (OSUG), Institut polytechnique de Grenoble - Grenoble Institute of Technology-Université Grenoble Alpes 2016-2019 (UGA), ECL, Corinne Leyval. Microbiogeochemistry of Cold Ecosystems. HAL CCSD; 2016

[56] Cowan DA, Cary SC, DiRuggiero J, Eckardt F, Ferrari B, Hopkins DW, et al. 'Follow the water': Microbial water acquisition in desert soils. Microorganisms. 2023;**11**:1670. DOI: 10.3390/microorganisms11071670

[57] O'Donnell M, Langston L, Stillman B. Principles and concepts of DNA replication in bacteria, archaea, and eukarya. Cold Spring Harbor Perspectives in Biology. 2013;**5**(7):a010108. DOI: 10.1101/cshperspect.a010108

[58] Dewachter L, Verstraeten N, Fauvart M, Michiels J. An integrative view of cell cycle control in *Escherichia coli*. FEMS Microbiology Reviews. 2018;**42**:116-136. DOI: 10.1093/femsre/fuy005

[59] Kelly B, Pearce EL. Amino assets: How amino acids support immunity. Cell Metabolism. 2020;**32**:154-175. DOI: 10.1016/j.cmet.2020.06.010

[60] Aliashkevich A, Alvarez L, Cava F. New insights into the mechanisms and biological roles of D-amino acids in complex eco-systems. Frontiers in Microbiology. 2018;**9**:683. DOI: 10.3389/fmicb.2018.00683

[61] Ma J, Sun Y, Gleichauf K, Lou J, Li Q. Nanostructure on taro leaves resists fouling by colloids and bacteria under submerged conditions. Langmuir. 2011;**27**:10035-10040. DOI: 10.1021/la2010024

[62] Jenkins J, Mantell J, Neal C, Gholinia A, Verkade P, Nobbs AH, et al. Antibacterial effects of nanopillar surfaces are mediated by cell impedance, penetration and induction of oxidative stress. Nature Communications. 2020;**11**:1626. DOI: 10.1038/s41467-020-15471-x

[63] Yang X, Zhang W, Qin X, Cui M, Guo Y, Wang T, et al. Recent progress on bioinspired antibacterial surfaces for biomedical application. Biomimetics. 2022;**7**:88. DOI: 10.3390/biomimetics7030088

[64] Baran A, Kwiatkowska A, Potocki L. Antibiotics and bacterial resistance—A short story of an endless arms race. International Journal of Molecular Sciences. 2023;**24**:5777. DOI: 10.3390/ijms24065777

[65] Zang J, Li M, Luo Q, Jin Y, Zhang L, Tang J, et al. Preparation and investigation of high-performance 3D Ag/SiO_2 nanohole array substrate based on cicada wing. Optical Materials (Amst). 2023;**136**:113412. DOI: 10.1016/j.optmat.2022.113412

[66] Fujimoto K, Saito A, Kotsuchibashi Y. Cicada-wing-inspired nanopillar hydrogels consisting of poly(vinyl alcohol) and poly(methacrylic acid) for capturing bacteria through their flexibility and wide range of motion. ACS Macro Letters. 2022;**11**:727-732. DOI: 10.1021/acsmacrolett.2c00126

[67] Elbourne A, Crawford RJ, Ivanova EP. Nano-structured antimicrobial surfaces: From nature to synthetic analogues. Journal of Colloid and Interface Science. 2017;**508**:603-616. DOI: 10.1016/j.jcis.2017.07.021

[68] Yang Y, Yang R, Fan C, Huang Y, Yan Y, Xu Y. Cicada wing-like hierarchical porous carbon to facilitate the re-deposition of sulfur for high-performance lithium-sulfur batteries. International Journal of Energy Research. 2022;**46**:23407-23417. DOI: 10.1002/er.8638

[69] Ivanova EP, Hasan J, Webb HK, Truong VK, Watson GS, Watson JA, et al. Natural bactericidal surfaces: Mechanical rupture of pseudomonas aeruginosa cells by cicada wings. Small.

2012;**8**:2489-2494. DOI: 10.1002/smll.201200528

[70] Wang J, Wang Q, Xu J, Li G, Ge J, Li Z, et al. $AlPO_4$ film with transparent anti-fogging function imitating cicada wings. Thin Solid Films. 2023;**781**:139996. DOI: 10.1016/j.tsf.2023.139996

[71] Cheng YW, Hsiao CW, Lin CH, Huang LY, Chen JS, Yang MC, et al. Bionic 3D periodic nanostructures by Ag nano-islands deposited on cicada wings for rapid SERS detection. Surface and Coating Technology. 2022;**436**:128323. DOI: 10.1016/j.surfcoat.2022.128323

[72] Tian G, Zhang Y, Feng X, Hu Y. Focus on bioinspired textured surfaces toward fluid drag reduction: Recent progresses and challenges. Advanced Engineering Materials. 2022;**24**:2100696. DOI: 10.1002/adem.202100696

[73] La Barre S, Bates SS. Marine biominerals with a biotechnological future. In: La Barre S, Bates SS, editors. Blue Biotechnology. New Jersey, USA: Wiley-VCH Verlag GmbH & Co. KGaA Hoboken; 2018. DOI: 10.1002/9783527801718.ch27

[74] Malshe A, Rajurkar K, Samant A, Hansen HN, Bapat S, Jiang W. Bio-inspired functional surfaces for advanced applications. CIRP Annals - Manufacturing Technology. 2013;**62**:607-628. DOI: 10.1016/j.cirp.2013.05.008

[75] Eadie L, Ghosh TK. Biomimicry in textiles: Past, present and potential. An overview. Journal of the Royal Society Interface. 2011;**8**:761-775. DOI: 10.1098/rsif.2010.0487

[76] Helbig R, Nickerl J, Neinhuis C, Werner C. Smart skin patterns protect springtails. PLoS One. 2011;**6**:e25105. DOI: 10.1371/journal.pone.0025105

[77] Dayan CB, Chun S, Krishna-Subbaiah N, Drotlef DM, Akolpoglu MB, Sitti M. 3D printing of elastomeric bioinspired complex adhesive microstructures. Advanced Materials. 2021;**33**:2103826. DOI: 10.1002/adma.202103826

[78] Yu C, Sasic S, Liu K, Salameh S, Ras RHA, van Ommen JR. Nature–inspired self–cleaning surfaces: Mechanisms, modelling, and manufacturing. Chemical Engineering Research and Design. 2020;**155**:48-65. DOI: 10.1016/j.cherd.2019.11.038

[79] Latthe SS, Terashima C, Nakata K, Fujishima A. Superhydrophobic surfaces developed by mimicking hierarchical surface morphology of lotus leaf. Molecules. 2014;**19**:4256-4283. DOI: 10.3390/molecules19044256

[80] Ensikat HJ, Ditsche-Kuru P, Neinhuis C, Barthlott W. Superhydrophobicity in perfection: The outstanding properties of the lotus leaf. Beilstein Journal of Nanotechnology. 2011;**2**:152-161. DOI: 10.3762/bjnano.2.19

[81] Zhao F, Zhan F, Wang L. Hybrid topography of lotus leaf under hydrostatic/hydrodynamic pressure. Advanced Materials Interfaces. 2023;**10**:2202044. DOI: 10.1002/admi.202202044

[82] Zhang X, Wang L, Levänen E. Superhydrophobic surfaces for the reduction of bacterial adhesion. RSC Advances. 2013;**3**:12003-12020. DOI: 10.1039/c3ra40497h

[83] Adak D, Bhattacharyya R, Barshilia HC. A state-of-the-art review on the multifunctional self-cleaning nanostructured coatings for PV panels, CSP mirrors and related solar devices. Renewable and Sustainable Energy

Reviews. 2022;**159**:112145. DOI: 10.1016/j.rser.2022.112145

[84] Upton RL, Crick CR, Crick CR. Pigmented self-cleaning coatings with enhanced UV resilience: Via the limitation of photocatalytic activity and its effects. Molecular Systems Design and Engineering. 2020;**5**:876-881. DOI: 10.1039/d0me00013b

[85] Rios PF, Dodiuk H, Kenig S, McCarthy S, Dotan A. Durable ultra-hydrophobic surfaces for self-cleaning applications. Polymers for Advanced Technologies. 2008;**19**:1684-1691. DOI: 10.1002/pat.1208

[86] Akhtar N, Thomas PJ, Svardal B, Almenningen S, De Jong E, Magnussen S, et al. Pillars or pancakes? Self-cleaning surfaces without coating. Nano Letters. 2018;**18**:7509-7514. DOI: 10.1021/acs.nanolett.8b02982

[87] Tuson HH, Weibel DB. Bacteria-surface interactions. Soft Matter. 2013;**9**:4368-4380. DOI: 10.1039/c3sm27705d

[88] Jaggessar A, Shahali H, Mathew A, Yarlagadda PKDV. Bio-mimicking nano and micro-structured surface fabrication for antibacterial properties in medical implants. Journal of Nanobiotechnology. 2017;**15**:64. DOI: 10.1186/s12951-017-0306-1

[89] Si Y, Dong Z, Jiang L. Bioinspired designs of superhydrophobic and superhydrophilic materials. ACS Central Science. 2018;**4**:1102-1112. DOI: 10.1021/acscentsci.8b00504

[90] Rasouli S, Rezaei N, Hamedi H, Zendehboudi S, Duan X. Superhydrophobic and superoleophilic membranes for oil-water separation application: A comprehensive review. Materials and Design. 2021;**204**:109599. DOI: 10.1016/j.matdes.2021.109599

[91] Shao H, Ma M, Wang Q, Yan T, Zhao B, Guo S, et al. Advances in the superhydrophilicity-modified titanium surfaces with antibacterial and pro-osteogenesis properties: A review. Frontiers in Bioengineering and Biotechnology. 2022;**10**:1000401. DOI: 10.3389/fbioe.2022.1000401

Chapter 5

Advanced Bioinspired Superhydrophobic Marine Antifouling Coatings

Mohamed S. Selim, Hamed I. Hamouda, Nesreen A. Fatthallah, Mohsen S. Mostafae, Shimaa A. Higazy, Samah Shabana, Ashraf M. EL-Saeed and Zhifeng Hao

Abstract

Following the tributyl-tin antifouling coatings' prohibition in 2003, global interest was directed toward non-toxic coatings as an eco-friendly alternative. Natural surfaces with superhydrophobicity exhibited exciting antifouling mechanisms. Efficient and eco-friendly antifouling coatings have been developed using bioinspired polymeric nanostructured composites. These superhydrophobic surfaces have rough topologies and low surface-free energies. Various organic/inorganic polymeric nanocomposites were developed for increasing fouling prevention by physical microfouling repulsion and chemical surface inertness. The biofouling costs and the difficulties of artificial antifouling coatings were also discussed in this chapter. It will introduce a cutting-edge research platform for next-generation antifouling surfaces for maritime navigation. This chapter aims to explain the evolution of superhydrophobic antifouling surfaces inspired by biological systems.

Keywords: antifouling coatings, environmentally friendly, self-cleaning nanocomposite coatings, polymer nanocomposites, nanofillers, bioinspired surfaces

1. Introduction

Modern research has focused on the development of effective marine antifouling and superhydrophobic surfaces to mitigate the detrimental effects of marine biofouling on the shipping sector [1]. Natural biomimetic structures with low surface-free energy (SFE) can provide inspiration for the creation of polymer/inorganic antifouling coatings. The key antifouling design tools required an understanding of the surface characteristics of nanocomposites, such as polymeric structures and nanofillers' morphology. Fouling represents a scary nightmare with dynamic issues that negatively affect marine navigation in both ecological and economic terms. Shipping contributes to almost 90% of all international trade. Shipping costs and environmental risks are just two of the negative ecological and economic effects of fouling adhesion [2]. Fouling increases fuel consumption and friction drag.

This can lower hydrodynamic efficiency and ultimately the ship's speed [3]. Increased CO2, NOx, and SOx emissions may result from fouling adherence (**Figure 1**), which raises shipping costs by around $12 billion annually [2, 3]. Traditionally, biocidal antifouling coatings were used to prevent biofouling. But such coatings have a severe impact on non-target species such as dolphins and fish, and shorten the time between dry-dockings [4]. Additionally, they may result in low fuel savings, environmental degradation, and harm to the maritime eco-system. In 2003, the International Maritime Organization (IMO) forbade the application of organo-tin antifouling paints due to these coatings' detrimental effects. Recent advancements have focused on FR coatings as an eco-friendly alternative to biocidal fouling-prevention substances.

Surfaces with FR coatings are non-toxic, very mobile, and non-stick. Fluoropolymers and polysiloxanes, two types of FR resin, have been shown to perform effectively as self-cleaning materials that can stop physical fouling through an anti-adhesion mechanism. Fluorine atoms are tightly bound in fluoropolymer structures, which results in high stiffness and a lack of structural rotation along the polymeric backbone [5]. This stiffness also makes it difficult for connected biofouling strains to easily free themselves. Polysiloxanes (particularly polydimethylsiloxane (PDMS)) offer a more widely used antifouling resin as compared to fluoropolymers [6].

PDMS building blocks include flexibility in the structure, non-toxicity, water-repellency, fouling-resistance, and reduced SFE [7]. Through the silicone matrix's well-dispersion of the nanofillers, the antifouling performance of PDMS was enhanced [8]. Superhydrophobic surfaces have made environmentally beneficial and cost-effective advancements in a variety of industries, including textiles, antifouling coatings, anti-icing, and anti-corrosion [9]. The ability of lotus leaves to self-clean is

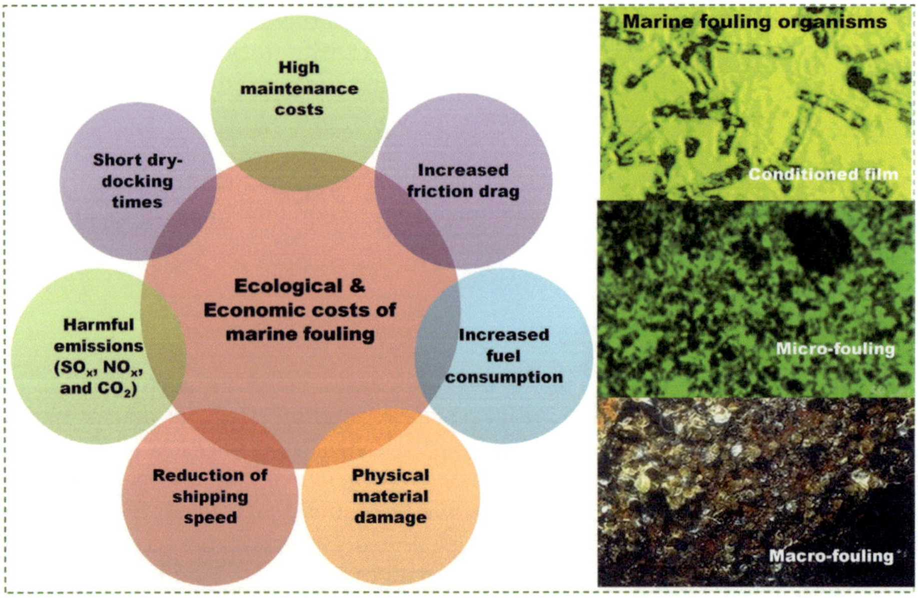

Figure 1.
The conditioned film, micro- and macro-fouling, and the issues brought on by fouling adherence on the ship hull are typical fouling phases that are graphically illustrated. An example of how fouling adhesion has a negative economic and ecological influence [1]. Copyright 2017; reprinted with Elsevier's consent.

one of the earliest lessons we can learn from nature. These leaves have been studied for decades because of their exceptional water repellency due to their distinctive structure and extremely hydrophobic composition [10]. The lotus leaves *Nelumbo nucifera* represent well-known examples of superhydrophobic surfaces that naturally clean themselves because of their 20–40 μm epicuticular wax rough topology [11]. Superhydrophobicity is found in nature in the form of butterfly wings, cabbage, Indian cress plant leaves, and more [12]. Due to their rough micro/nanostructure, low SFE, superhydrophobic surfaces have contact angle hysteresis ≤5° and water contact angles (WCA) ≥150° [13].

In 1990s, the concept of the first artificial superhydrophobic surface was demonstrated [14]. Sensitive surface features can be safeguarded by superhydrophobicity [15]. The Wenzel and Cassie-Baxter models can be used to explain how roughness affects a solid surface's non-wettability [16, 17]. The rise in the solid-liquid interface might increase surface roughness and hydrophobicity by trapping air in the surface grooves to produce superhydrophobic surfaces [18–20]. The resulting nanocomposite structures may have improved interfacial bonding due to economic, extra-high surface-area, and matrix-nanofiller interfacial bonding [21]. By evenly dispersing the nanofillers throughout the coating resin, it is possible to successfully build a superhydrophobic structure with outstanding substrate-coating adhesive forces. Rough topology superhydrophobic composites are possible when ceramic nanofillers and polymers are structurally controlled [22]. Superhydrophobicity can be produced by rough surfaces by trapping air in the ridges and grooves present beneath the water droplets. Biomimetic lithography, stamping, and etching processes were used to produce superhydrophobic nano-surfaces [23]. The creation of long-lasting antifouling nanostructures for the shipping industry was the main goal of the mature rostrum's nanocomposite coatings.

2. Detrimental effects of biofouling on shipping

2.1 Biofouling mechanism

Based on the biofoulans' size, type, and invasion time, biofouling (**Figure 1**) is commonly classified into three main categories: 1—first colonizers; 2—secondary; and 3—tertiary colonizers (soft and hard biofoulers) [24]. First and second colonizers are termed micro-biofoulers, whereas soft and hard biofoulers are generally termed macro-biofoulers [24, 25]. First colonizers are microorganisms that colonize invaded surfaces within hours. This microbial assemblage consists of bacteria, diatoms, unicellular microalgae, and cyanobacteria to form the so-called biofilm or slime [26].

In the early biofouling event (the first minutes), proteins, proteoglycans, and polysaccharides of different biological processes compose an organic-conditioning film consisting of chemical compound macromolecules to achieve wettable surfaces [27]. Planktonic bacteria, as one of the first colonizers, start the invasion through two stages within hours [28]. One of which is returnable, the so-called adsorption, and the other is unreturnable, the adhesion. The instant allure response caused by returnable adsorption attracts germs to the wet surface. Detrimental physical powers are known to control the bacterial adsorption process, including van-der-Waal forces, electrostatic interaction, and Brownian motion [29]. The so-called slime film, which is composed of a spreading bacterial lawn and extracellular polymeric bridges produced by sessile bacteria, is what forms the unreturnable phase.

Secondary colonizers (protozoans and macroalgal zoospores) start to emerge and colonize surfaces after the settlement of the first colonizers' stage within a period of almost a week [26, 30]. The primary colonizers provide them with enough nutrient supply *via* different microbial metabolic processes and products. A general insulation pattern of biofilm formation over time may be the net result of mutual and continuous physical interactions between the primary and secondary colonizers. Such physical interactions include conspiracies, survival tournaments, mutual elimination, and limited breakups.

Tertiary colonizers are macro-biofoulers (soft and hard biofoulers). Macro-biofoulers with non-calcareous structures are termed soft macro-biofoulers such as sponges and ascidians (tunicates) [31], whereas hard biofoulers are those with external calcareous skeletons, such as mussels, oysters, crustacean barnacles, and tube worms [32]. Tertiary colonization takes place after 2 to 3 weeks of settlement of both primary colonization and secondary colonization. However, the macroscopic tertiary biofoulers attach themselves to the invaded surfaces to feed on the primary and secondary colonizers [33]. Along the process of colonization, macro-biofoulers strongly engage themselves with the invaded surfaces *via* their secretions, which provide them with enough chemical bonding and electrostatic interactions to assure no breakup from such surfaces.

2.2 Marine biofouling costs

Biofouling assemblages cause ubiquitous calamity and negative economic and ecological impacts, especially in marine environment [34]. Maritime sectors appear to be adversely influenced by both micro- and macro-biofouling namely ships submerged surfaces, vessels and internal seawater systems, marine renewable energy, oil and gas, and aquaculture sectors [34–36]. In their colonization, macro-biofoulers

Maritime sector	Biofouling	Impact	Management
• Shipping industry • Oil & gas • Renewable energy • Aquaculture	Macro- and micro-biofoulers	• Reduction of propeller efficiency and increased surface roughness • Increased drag resistance and hydrodynamic weight • Decreased vessel's velocity loss of functionality and performance • Increased fuel consumption • Reduction of air quality and increased hazardous emissions	• Leaching antifouling biocides • Antifouling coatings (self-cleaning and foul-release)
	Micro-biofoulers	• Metallic corrosion (microbially induced corrosion)	

Table 1.
Major maritime biofouling-induced impacts and management [26, 35, 40, 41].

are responsible for an extra substantial economic cost because their hard-formed layers create typically optimum anaerobic conditions that accelerate the microbially induced metal corrosion process (anaerobic micro-biofoulers) [37, 38]. Hard macro-biofouling accounts for about 85% of annual economic impacts, whereas the total estimated penalties and fines of micro-biofouling and anaerobic microbially induced corrosion account for about 20 and 50% of all pipeline infrastructure failures [38]. The global estimation of the damage cost of microbially induced corrosion accounts for about 2.5 trillion USD annually [39]. Biofouling-induced damage costs are the net impacts of increased ship drag resistance, decreased velocity of the vessels, the hydrodynamic deficiencies of increased surface roughness due to the adhesion of biofouling communities, and, subsequently, exceeded biofouling management costs [40]. Major biofouling management regimes provide antifouling leaching biocidal coatings [38, 41], anti-adhesion (anti-biofilm) coatings [35], and self-cleaning foul-release coatings [35, 38]. Biofouling-induced impacts and management are summarized in **Table 1**.

3. Traditional methods to combat biofouling

The necessity to protect ship hulls against marine fouling invasion had been established for as long as ships are needed for transportation. The tale of antifouling paints started in ancient time with the complaints of different sailors with their wooden ship hulls from the heavy burden of various assemblages of marine organisms and seaweeds on the submerged surfaces. Such accumulation negatively affected the vessels as it affected the overall ship's performance [42]. Such suffering forced the sailors to provide their wooden ships with different coverings such as asphalt, wax, tar, oil, tallow [43, 44]. In the ages of 700 B.C., the Phoenician and Carthaginian sailors began to apply pitch-copper and tallow-lead thin sheets as protective shields from the marine invaders [45]. In the ages after 500 B.C., Phoenicians applied oil-mixed arsenic and sulfur sheathing. Protection by asphalt, wax, tar, tallow, pitch, copper, and lead had been applied during the ages from 300 B.C. until the fifteenth century among Greeks, Romans, Vikings, and Columbus sailors [46]. In the early 50's, organometallic paints such as tin, arsenic, and mercury started to emerge and encountered a development to introduce tributyltin (TBT)-based paints that showed amazing performance [23]. Later on, in 2001, TBT was banned by the IMO due to the severe toxicity to the marine environment [47]. During the period between 1960's–1970's and the early nineteenth century, copper sheathing of wood and steel ships had been commonly used and favored with different mixtures of copper sulfide, arsenic ore, iron dust, and zinc alloy. From the late period of 1970's–1980's to the mid-nineteenth century paints containing copper, zinc, and mercuric oxide had been applied with polymeric binder [36].

Considering the foregoing history, antifouling paints could be classified into different categories: 1—biocidal paints (as organotin compounds such as mercury, arsenic, and their compounds), 2—leaching paints (containing polymers with high molecular weight, such as epoxy, acrylates, and chlorinated rubber), 3—ablative paints (a mixture of biocide, such as iron, zinc oxide, arsenic, and mercury, incorporated with the coating resin), 4—self-polishing paints (acrylic or methacrylic copolymer-based paints such as TBT-based paints and tin-free paints based on copper, zinc, and silyl acrylate), 5—foul-release paints (their mode of action depends on prevention fouling adhesion *via* their smooth and low-energy surfaces that enable the hydrodynamic forces of water to wash off the fouling attachments, where the paint is based on the elastomers PDMS, the epoxy primer, and the oil additive to enhance the smooth performance), 6—engineered antifouling surfaces (physically, chemically, and topographically controlled surfaces inspired by different natural antifouling surfaces found around us in nature, such as shark and whale skin and molluscan shells).

4. Innovative approaches, such as superhydrophobic surfaces, to tackle biofouling

4.1 Eco-friendly FR coatings

Fouling refers to the accumulation of marine organisms, such as algae, barnacles, and mussels, on underwater surfaces such as ship hulls, offshore structures, and marine equipment [48–50]. This fouling can lead to increased drag, reduced fuel efficiency, and increased maintenance costs. The rise in fuel consumption inherently results in elevated transportation expenses for the shipping sector. About 2% of the world's total energy is allocated to commercial shipping, encompassing a fleet of nearly 100,000 cargo vessels [51]. Consequently, the excessive fuel usage attributed to biofouling can lead to heightened CO_2 emissions, which is counterproductive in the pursuit of carbon neutrality. Furthermore, organisms adhering to vessel hulls can traverse international waters, reaching local marine ecosystems devoid of their natural predators. In such settings, they can perpetuate unchecked and potentially trigger ecological upheavals [52]. Current data indicates that governments and enterprises on a global scale expend over $15 billion each year to mitigate or manage the impacts of marine biofouling [53].

Traditional antifouling coatings typically contain toxic biocides, such as TBT, which was prohibited by the IMO in 2001 [54, 55], quaternary ammonium salts, tar, asphalt, toxic (arsenic), or heavy metals (silver, copper, and lead), which can have negative impacts on marine ecosystems [56–58]. Eco-friendly FR coatings can effectively prevent the attachment of marine organisms (such as algae, barnacles, and mollusks) to submerged surfaces while minimizing environmental impacts [59, 60]. Eco-friendly fouling-release coatings can provide an effective alternative to traditional antifouling coatings. These coatings work by creating a low-friction surface, preventing marine organisms from firmly attaching to the submerged surface. The reduced friction allows water flow to easily remove any loosely attached organisms, preventing the formation of a thick fouling layer. Eco-friendly FR coatings are designed to be non-toxic and environmentally friendly. They are typically made from non-biocidal materials, such as silicone or fluoropolymer-based polymers, which have low toxicity and minimal impact on marine life [61, 62]. These coatings are engineered to have a smooth and slippery surface, making it difficult for fouling organisms to establish a strong attachment. Eco-friendly FR coatings can provide effective fouling prevention while minimizing negative environmental impacts. By reducing the need for toxic biocides, these coatings can help protect marine ecosystems and contribute to sustainable marine practices [63, 64]. Recent developments are based on FR coatings, including silicone, modified fluorinated polymers, cross-linked coatings, amphiphilic copolymer coatings, hydrogel coatings, and biomimetic coatings [65, 66]. Additionally, environmentally friendly compositions comprising curable polysiloxane binders and non-ionic hydrophilic-modified polysiloxanes have been utilized [67, 68].

For example, a recent paper highlights the importance of designing coatings that are bactericidal to prevent fouling by bacteria, marine organisms, and proteins. It discusses the major strategies for combating surface fouling, including preventing biofoulants from attaching or degrading them, and the use of various techniques such as functionalization of surfaces with poly (ethylene glycol) or oligo (ethylene glycol), biocidal agents, polycations, enzymes, nanomaterials, and photoactive agents [69]. Various innovative technologies have been employed to improve these coatings, including the use of copolymers containing MQ silicone and acrylate [65, 70], charcoal-based graphene oxide-copper oxide [71], and the incorporation of phenylmethyl silicone oil as a lubricant [72].

The main problem of biofouling is the accumulation of marine organisms on submerged surfaces and its negative impacts on various industries, such as shipping, aquaculture, and energy production [73–75]. There is a need for effective antifouling coatings that can prevent or reduce biofouling, and for discussing the limitations of current coatings, such as toxicity, environmental concerns, and durability. This work introduces the concept of FR coatings, which prevent biofouling by reducing the adhesion strength of marine organisms to the surface Also, it describes the advantages and challenges of these coatings. Overall, the introduction sets the context and motivation for the research presented in the paper and highlights the importance of developing new FR coatings based on innovative materials and formulations. The paper discusses the development of FR coatings based on copolymers containing MQ silicone and acrylate, and the preparation of FR paints based on these copolymers with the incorporation of non-reactive phenylmethylsilicone oil [70]. The coatings showed good hydrophobicity, mechanical properties, and adhesion strength, with high oil leaching efficiency and excellent antifouling performance. The paper also provides insights into the effects of copolymer composition on the properties of the coatings, which can guide the design of new coatings with tailored properties.

It is important to note that the effectiveness of FR coatings can vary depending on factors such as the specific coating formulation, water temperature, vessel speed, specific marine environmental conditions, and the type of fouling organisms present. Ongoing research and development are crucial to improving the performance and longevity of these eco-friendly coatings. Further studies and field trials are necessary to evaluate the long-term effectiveness, durability, and economic viability of eco-friendly FR coatings. However, the hypothesis suggests that these coatings hold promise as a more sustainable alternative to traditional antifouling coatings, benefiting both marine industries and the environment. FR coatings work on the principle of creating a slippery or low-energy surface that makes it difficult for organisms to adhere to the substrate.

4.2 Bioinspired strategies for antifouling applications

Biofouling is a complicated, dynamic phenomenon that affects both the environment and the economy on a worldwide scale. Through collaborative interdisciplinary efforts in marine biology, polymer science, and engineering, bio-inspired tactics have been harnessed to pave the way for advancing the next era of antifouling marine coatings [76, 77]. Bioinspired antifouling strategies aim to reduce the environmental impact of conventional antifouling coatings by utilizing sustainable and nature-inspired solutions. However, it is important to note that while these approaches hold promise, they may have limitations in certain environments or applications. Ongoing research and development are focused on optimizing the performance and durability of bioinspired antifouling coatings. There have been several bioinspired antifouling coatings that have been successfully developed and tested. These coatings draw inspiration from nature's strategies to prevent fouling and have shown promise in reducing or preventing the attachment of marine organisms [76–79].

Adhering to the principles of nature and drawing insights are the natural world form the foundational principles of bionic technology. Within this framework, antifouling approaches can be categorized into two main domains: chemical and mechanical. (i) Numerous aquatic and terrestrial organisms possess the ability to secrete natural substances that combat bacterial infections and biofouling [78, 79]. As an illustration, terpenoid and halogenated compounds sourced from algae serve as notable examples, and capsaicin produced from chili peppers [80, 81], also has antifouling action [82]. (ii) Mechanical antifouling approaches: Within this realm, scientists have identified intrinsic surface textures spanning from nanometer to millimeter dimensions across various organisms, including mangrove foliage, lotus leaves (*Nelumbo nucifera*), shells, sharkskin, and corals [83]. These natural topographies operate mechanically to diminish the prevalence of biofouling [84, 85]. The main bioinspired approach is the surface roughness and texture of marine organisms that have evolved rough or textured surfaces that discourage fouling attachment. Bioinspired antifouling coatings mimic these features to create surfaces that are less favorable for the attachment and growth of fouling organisms.

Also, the microtopography of some organisms that have tiny ridges and scales called denticles that reduce drag and discourage fouling, such as shark skin [84, 85]. Coatings with microtopography can mimic these structures to create a slippery surface that makes it difficult for organisms to settle and grow. These coatings, known as "sharkskin coatings" or "riblet coatings," have been applied to ship hulls and have shown promising results in reducing drag and fouling attachment.

Schumacher et al. [86, 87] successfully recreated shark epidermal riblets on a PDMS elastomer to establish a microstructured surface. The architecture featured rib structures forming periodic rhombuses with dimensions of 2 mm width, lengths ranging from 4 to 16 mm, and 2-mm spacing. Notably, the outcomes revealed a significant decline in spore settlement on the engineered surface in contrast to smooth surfaces, manifesting an impressive effective reduction of up to 77%. Moreover, the surface exhibited the capability to minimize bacterial adhesion and suppress the formation of biofilms. Thus, these shark-inspired surfaces exhibit the potential to proficiently counteract microbial sedimentation and impede biofilm establishment.

Some lotus leaves and other plants have self-cleaning properties due to their waxes and hydrophobic surfaces with micro−/nano-structures that repel water and contaminants [88, 89]. To date, these principles are being applied to create self-cleaning antifouling coatings that shed water and debris, reducing the attachment of organisms [90]. As an illustration, Selim and colleagues effectively prepared a range of nanostructured coatings with a lotus-inspired configuration. This innovative approach involved utilizing *in situ* techniques to combine vinyl-terminated PDMS with $YeAl_2O_3$ nanorod composites. These composites, composed of single-crystal nanorods, are easily synthesized and possess regulated structures and shapes [91, 92]. Consequently, the resultant surface inherits superhydrophobic attributes, self-cleaning attributes, and noteworthy antifouling properties.

Furthermore, certain marine organisms use hydrophobic (water-repellent) or hydrophilic (water-attracting) surfaces, produce natural anti-adhesives or compounds, produce protective mucus layers, or can change their surface properties in response to environmental cues to control fouling. Some marine organisms have biocidal activity and produce compounds that are toxic to fouling organisms. Researchers are studying these properties and compounds to develop environmentally friendly antifouling coatings that mimic their effects, create a physical barrier, and can switch between fouling-resistant and fouling-release states, deterring fouling attachment or making it easier to clean. Additionally, researchers are exploring the use of naturally occurring substances such as enzymes and proteins to create antifouling coatings that interfere with the attachment and growth of fouling organisms.

Unlike mobile species like sharks and dolphins that can actively remove attached fouling organisms, sedentary species such as sponges and corals lack this ability. However, despite their stationary nature, corals manage to maintain clean and tidy skin surfaces. Certain sponges and corals have intricate structures that create turbulent water flow, preventing fouling attachment. Coatings inspired by these macrotextures have been developed, creating surfaces with complex patterns and features that make it difficult for fouling organisms to attach. For example, previous research has demonstrated that corals employ a variety of antifouling mechanisms [91, 92]. These strategies include the release of natural antifoulants, the utilization of skin with low surface energy, the emission of fluorescence, the shedding of old skin layers, and the swinging tentacles. Last but not least, biomimetic surface coatings directly replicate the surface features and properties of specific marine organisms known for their fouling resistance.

4.2.1 Superhydrophobicity in nature

During the 1990s, researchers in the fields of biology and materials science initiated investigations into naturally occurring superhydrophobic surfaces [93]. Over the past few years, scientists have introduced superhydrophobic coatings characterized by remarkable water repellency for automotive, marine, and medical sectors [94, 95]. The emergence of superhydrophobic nanocoatings represents a pivotal approach to achieving surface superhydrophobization, leading to remarkable alterations in both surface and interfacial domains. Numerous techniques have been established to create superhydrophobic surfaces, encompassing approaches such as layer-by-layer assembly, sol-gel method, photolithography, electrodeposition, electrospinning, and 3D printing [96]. While multiple methodologies exist, those founded on simple chemical reactions hold particular appeal due to their straightforward procedures, economical nature, and potential for large-scale production.

On a superhydrophobic surface, the water droplet forms a nearly spherical shape with minimal contact area, and it tends to slide off the surface easily due to the combination of surface texture and low surface energy [97]. A superhydrophobic surface is characterized by a high water contact angle (WCA) exceeding 150° [98, 99]. Superhydrophobicity, or the ability to repel water and maintain extremely high contact angles with water droplets, is observed in various natural phenomena. Here are a few examples of superhydrophobicity in nature:

An illustrative instance of superhydrophobicity is the surface of a lotus leaf. Lotus leaves boast remarkable water repellency and self-cleaning capability due to the presence of hierarchical micro−/nanostructures and hydrophobic waxes on their surfaces, a phenomenon referred to as the “lotus effect.” Insect wings, such as those of Morpho butterflies, similarly demonstrate characteristics of the “lotus effect” [93]. The wing surfaces feature intricate micro- and nanostructures that scatter light, giving the wings their vibrant colors. These structures also create a hydrophobic surface, causing water droplets to bead up and roll off the wings. Additionally, water striders are insects that can walk on the water surface due to their hydrophobic legs. The legs are covered with tiny hairs that trap air and create a cushioning effect, preventing the insect from breaking the water's surface tension. This adaptation allows water striders to move effortlessly on the water without sinking.

Selim et al. introduced a self-cleaning solution utilizing elastomeric PDMS and nanomagnetite composites [100]. The composite surface demonstrated superior antifouling efficacy compared to the PDMS, PDMS/nanomagnetite, and control composites, as observed through diatom and bacterial fouling assessments. Furthermore, Selim and his colleagues explored the effects of nanostructured surfaces and the distribution of GO decorated with TiO_2 nanorod (NR) fillers on the FR and superhydrophobic properties of silicone nanocoatings [101]. By integrating 1 wt% GO/TiO_2 hybrid fillers, they achieved enhancements in surface roughness and non-wettability while reducing surface-free energy. Consequently, they developed an improved model for superhydrophobic nanocomposite coatings, enabling sustained surface attributes suitable for prolonged marine applications. These examples from nature inspire the development of superhydrophobic coatings and materials in extensive applications, such as drag reduction, antifouling, self-cleaning surfaces, anti-icing coatings, oil/water separation, anticorrosion purposes, and waterproof textiles. By emulating the structures and properties found in these natural systems, scientists and engineers aim to create functional and efficient superhydrophobic materials for a range of practical applications [102–105].

4.2.2 Characterization of the superhydrophobic surfaces

The fabricated superhydrophobic materials are subjected to various tests to evaluate their hydrophobic properties. Contact angle measurements are commonly used to quantify the degree of water repellency, with higher contact angles indicating greater hydrophobicity. Other tests may include nanoscale roughness, water droplet roll-off angles, coating thickness, durability assessments, mechanical stability, and resistance to contamination [106].

The water contact angle (WCA) measures the angle between a water droplet and the surface. A WCA greater than 150° indicates superhydrophobicity. Several factors, including surface roughness, surface energy, and the level of cleanliness, influence the contact angle [107, 108]. When a liquid effectively wets the surface (referred to as a wetting liquid or hydrophilic surface), the static contact angle assumes a value within the range of $0^0 \leq \theta \leq 90^0$ degrees. Conversely, when a liquid does not wet the surface (termed a hydrophobic surface), the contact angle falls within the range of $90^0 < \theta \leq 180^0$ degrees (**Figure 2**).

In 1805, Thomas Young introduced the foundational concepts of contact angle and wettability [109]. Presently, the contact angle measurement technique stands as the most prevalent approach employed to assess the wettability of a surface [110, 111]. The static contact angle (θ) is defined as the angle formed at the point where the gas-liquid interface intersects the three-phase boundary, represented by a tangent line.

The Wenzel state manifests when the interactions between the liquid and the solid are high, resulting in the complete filling of surface voids by the liquid droplet [112]. Meanwhile, when the voids are occupied by air, the liquid droplet resides on both the solid surface and the air. This state is called the Cassie-Baxter state. In contrast to Wenzel's model, the Cassie-Baxter model exhibits a slightly reduced contact angle

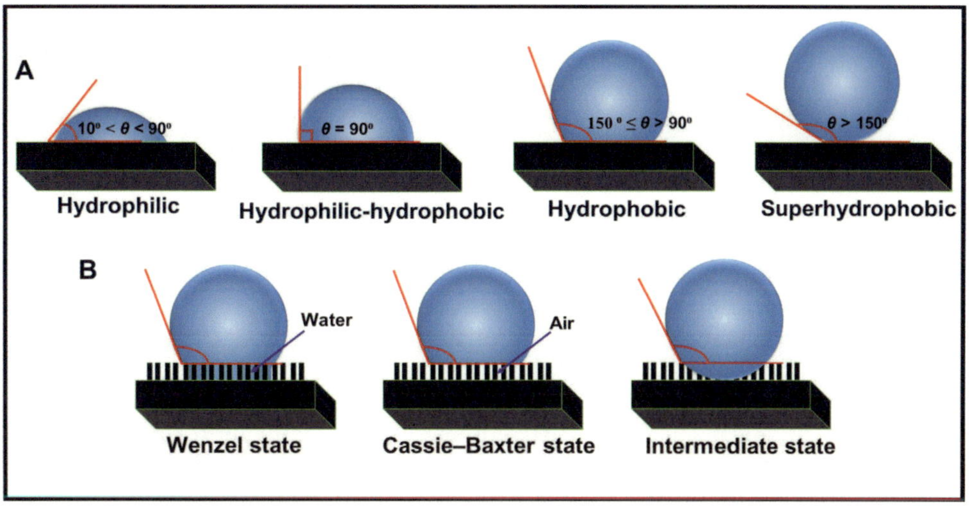

Figure 2.
(a) Example of the contact angles created by sessile liquid drops on a uniformly flat solid surface; (b) a droplet's non-wetting behavior on solid substrates, including Wenzel's and Cassie-Baxter models.

hysteresis and an increased advancing contact angle. Consequently, the Cassie-Baxter model is more extensively employed across various surfaces [113–115].

4.3 Bioinspired antifouling nanocomposite coatings

Numerous plants and animals have evolved surfaces with exceptional antifouling performance in nature [85, 116–118]. Strong attempts have been made to create coatings that resemble these natural antifouling surfaces because they serve as inspiration. In recent years, six main bioinspired strategies have been used to tackle biofouling: (i) micro–/nano-rough coatings: Some creatures, including the lotus and shark [85, 118], have developed micro/nanostructured surfaces to resist biofouling. (ii) Natural antifouling agents: One effective method to stop biofouling is to employ bioactive substances that are emitted by particular organisms, like sponges and corals [119]. Hydrogels resemble mucus. In other cases, fish and amphibians (like frogs) use the epidermal mucus to counter biofouling [120]. (iii) Slippery liquid-infused porous surfaces (SLIPS): Microstructures can be found on the skin of earthworms and pitcher plants [121]. They differ from the lotus, though, in that liquid is present and trapped in the microstructures. These interfaces, known as SLIPS, are made to be nonstick and can therefore be employed to fend off fouling organisms [122]. (iv) Some marine creatures molt epithelium to refresh their surface [123], which aids in clearing their surfaces of fouling organisms. Dolphin skin has a rubber-like feel and can generate an unsteady surface when subjected to turbulent flow [124, 125]. Fouling organisms have a hard time settling on a surface that is so unstable. Because of the shedding effect and the unstable surface, the surface is dynamic in marine environments. (v) Zwitterionic coatings: Bilayers of lipids in living things contain phosphotidylcholine units, which have zwitterions made up of two organisms with opposing charges. Researchers investigated the antifouling potential of phosphatidylcholine's analogues and specifically created zwitterionic polymers, as a result of their blood clotting-preventing capabilities [85, 126, 127].

These bioinspired techniques exhibit none of TBT's harmful environmental effects and have excellent antifouling properties; therefore, they have great potential for use in the future. The main bioinspired tactics used to prevent biofouling are highlighted in this chapter, which also provides new and intriguing results from nature. These bioinspired methodologies are also discussed in terms of their cutting-edge materials, production methods, pertinent antifouling processes, current issues, and potential future developments. The advantages of nanomaterials include their high surface area-to-volume ratio and economic advantage. The mechanical robustness and surface non-wettability are improved thanks to nanofillers' promotion of mechanical strength and roughness. Because nanofillers typically have inherent antifouling activity and because rough surfaces will entrap air more than smooth surfaces, they can enhance the antifouling characteristics of surfaces [128, 129].

To create these antifouling coatings, a variety of nanofillers, including Cu_2O [130], Ag [131], MnO_2 [132], and graphene oxide [133, 134] were added to polymeric surfaces. These coatings are affordable, effective, lasting, harmless, and friendly to the environment. It is becoming more common to create bioinspired antifouling coatings using nanocomposite methods for innovative and large-scale production. Natural antifoulants have been shown to have a number of antifouling mechanisms; however, there is insufficient information to definitively pinpoint the molecular pathways. Natural antifoulants' mechanisms are still poorly understood [135]. There is still much to learn about how natural antifoulants work to prevent fouling.

4.4 Marine antifouling biomimetic surface microtopographies

Numerous maritime animals possess fouling-resistance qualities, which were affected by the surface's texture roughness, wettability, surface energy, etc. [136, 137]. Since no biocides are released, there is less potential danger to marine ecosystems when duplicating the natural antifouling surface microstructures. So, one of the upcoming research trends is biomimetic microstructure coating, which has received a lot of attentions lately [138, 139]. Rough microstructures may have an antifouling mechanism that reduces the surface area where fouling organisms come into contact with the surface [140, 141]. Surface roughness has a stronger fouling-resistance [142, 143]. When the structure is an order of magnitude smaller than the fouling organism, the number of attached foulers grows but adhesion strength decreases [144]. Gorgonians, shark skin, echinoderms, and other biomimicking antifouling surfaces were examined by Scardino and de Nys [145]. Terrestrial wildlife also offers priceless models, so inspiration is not confined to marine critters.

A number of physical fabrication techniques, such as picosecond laser texturing, electron beam lithography, hot embossing, ion beam lithography, photolithography, and soft lithography, were evaluated in order to replicate the biological surface microtopographies [146, 147]. Soft mold replication is the most common and straightforward method for recreating biological microstructures on surfaces coated with synthetic materials. The elastomer sheet converts into a negative replica of the biological sample after curing and is removed from the mold. The negative replica is then secured to a plate and used as a mold. The specimen's positive replica is obtained after curing and separation [148]. The negative replica is covered with the precursor and curing agent. On the surface of lotus leaves, waxes and microstructures have been discovered [149–152]. A layer of air can be trapped between a drop of liquid and the surface of a leaf by the hydrophobic epicuticular waxes and microstructures (Cassie-Baxter state). Both characteristics produce surface self-cleaning on lotus leaves with low CAH and stop moisture from seeping into the surface [153, 154]. Every micropapillary on the surface of the lotus leaf was found to be covered in nanostructures [155]. The micro–/nano-hierarchical composite architectures allow lotus leaves to be extremely hydrophobic and low sticky [156, 157]. When used in marine vessels, the microstructures can therefore have a dual effect.

Numerous fabrication processes have been used to create micro–/nanostructured surfaces thanks to the development of cutting-edge technologies [158]. These techniques include additive manufacturing (3D printing) [159], mechanical micromachining [160], self-assembly [161], etching procedures [162], electrostatic approaches [163], nanocomposite approaches [144], deposition approaches [164], and soft lithography [164]. Although there may be certain disadvantages, these techniques have a bright future when used in actual manufacturing. A successful FR coating of PDMS/β-SiC nanocomposite with 153° WCA and micro-roughness was prepared (**Figure 3**) [165]. Additionally, a silicone/ZnO nanostructured composite with micro/nano-roughness and surface superhydrophobicity was developed (**Figure 4**) [166]. PDMS enhanced with TiO_2/SiO_2 nanocomposite was created [167] for photocatalytic and self-cleaning outcomes. AFM examination of the produced well-dispersed silicone/$TiO_2@SiO_2$ (0.5 wt.%) indicated surface roughness and displayed superhydrophobic antifouling performance.

Improved superhydrophobicity and fouling-inhibition properties may also be a result of this nano–/micro-surface and low SFE [112, 168]. Two recently created nanocomposites of PDMS/RGO and PDMS/GO-γ-AlOOH coating were compared as superhydrophobic FR materials [169]. Compared to PDMS/RGO nanocomposite, the nanocoating of PDMS/GO-γ-AlOOH (3 wt.%) composite displayed better FR

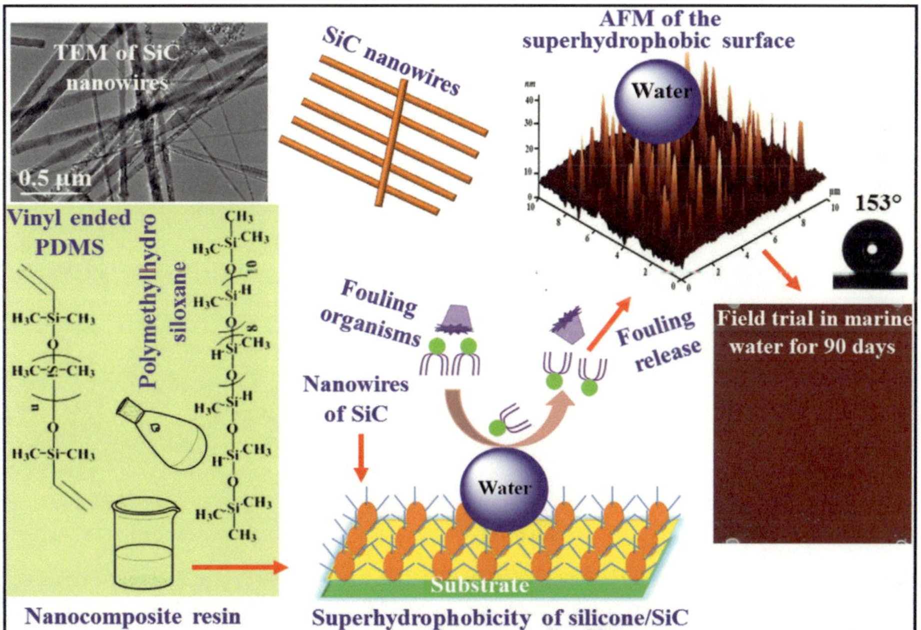

Figure 3.
Hydrosilation-cured silicone/β-SiC nanostructured composite as antifouling and superhydrophobic surface. The TEM of β-SiC nanowires and the AFM of the FR-causing PDMS/β-SiC composite are also included. The created nanocomposite underwent a 3-month field trial in seawater, and the results showed excellent fouling-prevention [165]. Copyright 2018 was adopted with springer Nature's consent.

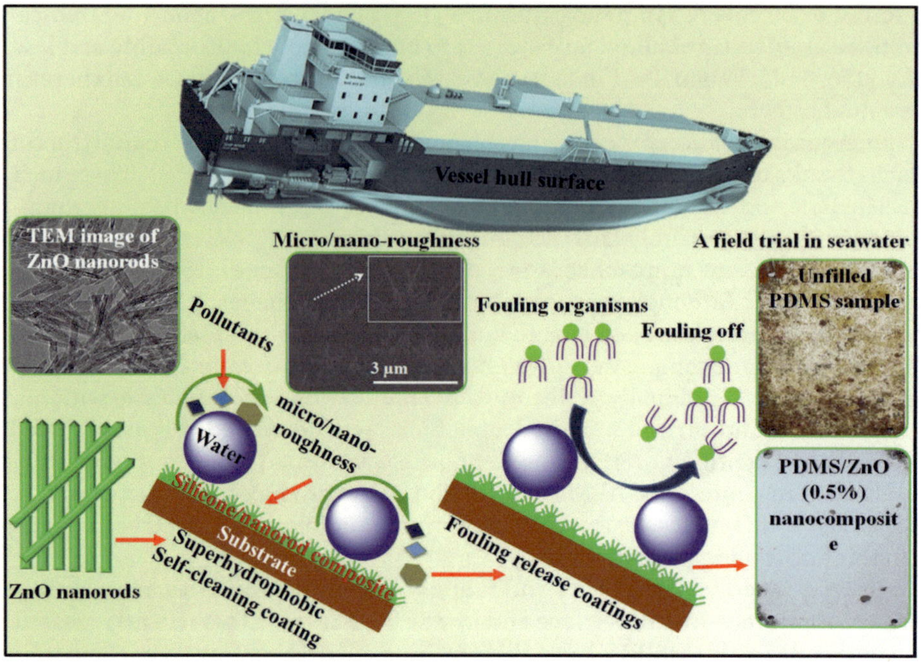

Figure 4.
Illustration of the silicone/ZnO nanostructured composite film [166] (copyright 2019, after permission from Elsevier).

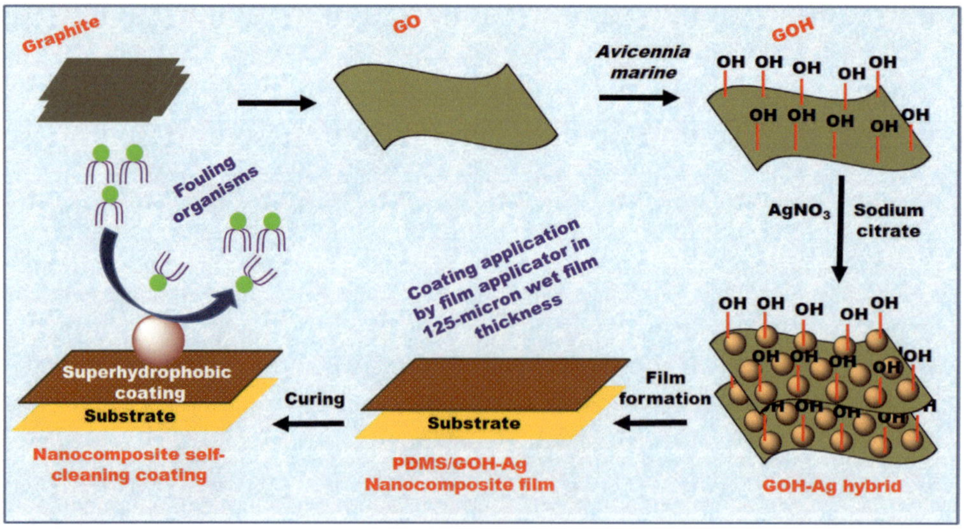

Figure 5.
The PDMS/GOH@Ag (0.5 wt.%) nanofillers for antifouling and FR nanocoatings [170]. Copyright 2021, reproduced with permission from Elsevier.

performance under various microfouling stressors. The well-dispersed silicone/GO-γ-AlOOH (3 wt.%) could achieve the lowest biodegradability of 1.6%. It has microbial endurability against gram-negative and gram-positive bacterial strains with 97.94%, and 86.42%, respectively.

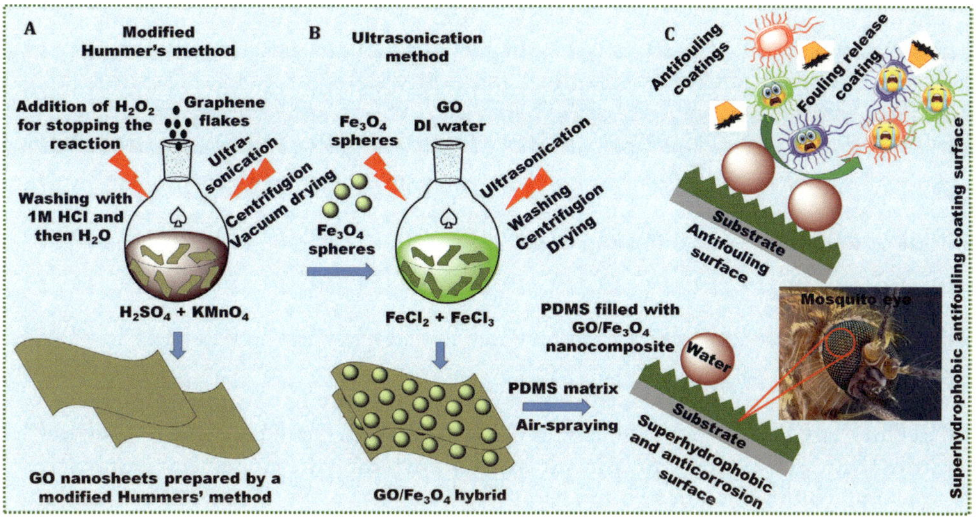

Figure 6.
Schematic illustration for the PDMS/GO-Fe_3O_4 (1 wt.% nanofillers) for superhydrophobic antifouling nanocomposite coatings [171]. Copyright 2023, reprinted with Elsevier's permission.

Avicennia marina/silver was employed to reduce GO and obtain GOH@Ag hybrid as reported by Soleimani et al. [170] (**Figure 5**). In PDMS-based coatings, GO, GOH@Ag, and CNTs were all utilized as nanofillers. The GOH@Ag (0.5 wt.%) showed the greatest performance with SFE = 16 mN/m, root mean square roughness (RMS) = 103 nm, and WCA = 118.8°. The production of PDMS/GOH@Ag nanocomposite was achieved by the combination of Ag and *A. marina* to the graphene nanocomposite coatings.

A surface with a micro−/nano-roughness, low SFE (12.06 mN/m), high WCA (158° ± 2°), and high FR efficiency was created by PDMS/GO-Fe_3O_4 (1 wt.%) nanocomposite as reported by Selim et al. [171] (**Figure 6**). It showed the lowest biodegradability percentage when tested against *Kocuria rhizophila* (2.047%), *Pseudomonas aeruginosa* (1.961%), and *Candida albicans* (1.924%).

5. Conclusions

The intriguing characteristics of naturally occurring superhydrophobic surfaces have inspired the creation of new superhydrophobic antifouling nanocoatings. Surface non-wettability and FR structure were significantly influenced by surface heterogeneity and rough topology. Graphene is a practical substance for producing superhydrophobic surfaces due to its distinct physicochemical features. Water-repellent antifouling marine coatings have gained much interest for their advantages in both the economy and the environment. Numerous characteristics of graphene nanocomposites, including antifouling, self-cleaning, and anti-corrosion, are available at a low price. Graphene nanostructured surfaces have outstanding antifouling and anticorrosion features. The potential of nanoscale filler tectonics was illustrated for the fabrication of structurally folded nanocomposite materials. Self-cleaning and antifouling graphene nanocomposites have been produced and are being used as a new trend in antifouling applications in marine navigation. A variety of hybrid graphene composite materials are being researched for maritime coatings due to their superior surface properties. Well-dispersed graphene-based nanocoatings can exhibit outstanding physico-mechanical and non-wetting properties. Benefits of these nanocomposites include coating behavior that prevents fouling, facility, economics, and toughness. The fabrication of PDMS/nanofiller composites can result in FR nanostructures with improved self-cleaning capabilities. These multifunctional FR nanocomposites will be commercially available in the near future, saving tens of billions of dollars in annual fouling costs.

Acknowledgements

This work was supported by the Open Fund of the Key Laboratory of Science and Technology for National Defence (China) and the Egyptian Petroleum Research Institute (EPRI), Cairo, Egypt. Also, it was supported by the financial support from the National Natural Science Foundation of China (Grant No: 52150410401) and the post-doctoral research fellowships of Dalian Institute of Chemical Physics, Chinese Academy of Sciences, CAS, Dalian, 116023, China.

Declaration of competing interest

The authors declared that they have no conflict of interest.

Author details

Mohamed S. Selim[1,2*], Hamed I. Hamouda[3,4], Nesreen A. Fatthallah[4], Mohsen S. Mostafae[5,6], Shimaa A. Higazy[2], Samah Shabana[7,8], Ashraf M. EL-Saeed[2] and Zhifeng Hao[1*]

1 Guangdong Provincial Key Laboratory of Plant Resources Biorefinery, School of Chemical Engineering and Light Industry, Guangdong University of Technology, Guangzhou, PR China

2 Petroleum Application Department, Egyptian Petroleum Research Institute (EPRI), Cairo, Egypt

3 Dalian Engineering Research Center for Carbohydrate Agricultural Preparations, Liaoning Provincial Key Laboratory of Carbohydrates, Dalian Institute of Chemical Physics, Chinese Academy of Sciences, CAS, Dalian, PR China

4 Processes Development Department, Egyptian Petroleum Research Institute (EPRI), Nasr City, Cairo, Egypt

5 Refining Department, Egyptian Petroleum Research Institute, Nasr City, Cairo, Egypt

6 CAS Key Laboratory of Standardization and Measurement for Nanotechnology, CAS Center for Excellence in Nanoscience, National Center for Nanoscience and Technology, Beijing, PR China

7 Department of Biochemistry and Molecular Biology, College of Marine Life Sciences, Ocean University of China, Qingdao, PR China

8 The Egyptian Ministry of Health and Population, Cairo, Egypt

*Address all correspondence to: moh.selim_chem2006@yahoo.com, mohamedselim@gudt.edu.cn and haozf@gdut.edu.cn

References

[1] Selim MS, Shenashen MA, El-Safty SA, Sakai M, Higazy SA, Selim MM, et al. Recent progress in marine foul-release polymeric nanocomposite coatings. Progress in Materials Science. 2017;**87**:1-32

[2] Oliveira DR, Granhag L. Ship hull in-water cleaning and its effects on fouling-control coatings. Biofouling. 2020;**36**(3):332

[3] Hu P, Xie Q, Ma C, Zhang G. Silicone-based fouling-release coatings for marine antifouling. Langmuir. 2020;**36**(9):2170

[4] Ciriminna R, Bright FV, Pagliaro M. Ecofriendly antifouling marine coatings. ACS Sustainable Chemistry & Engineering. 2015;**3**(4):559

[5] Wang J, Jin X, Li C, Wang W, Wu H, Guo S. Graphene and graphene derivatives toughening polymers: Toward high toughness and strength. Chemical Engineering Journal. 2019;**370**:831

[6] Maan AMC, Hofman AH, de Vos WM, Kamperman M. Recent developments and practical feasibility of polymer-based antifouling coatings. Advanced Functional Materials. 2020;**30**(32):2000936

[7] Selim MS, Elmarakbi A, Azzam AM, Shenashen MA, EL-Saeed AM, El-Safty SA. Ecofriendly design of superhydrophobic nano-magnetite/silicone composites for marine foul-release paints. Progress in Organic Coating. 2018;**116**:21-34

[8] Selim MS, Yang H, Wang FQ, Li X, Huang Y, Fatthallah NA. Silicone/Ag@SiO_2 core–shell nanocomposite as a self-cleaning antifouling coating material. RSC Advances. 2018;**8**:9910-9921

[9] Elbourne A, Crawford RJ, Ivanova EP. Nano-structured antimicrobial surfaces: From nature to synthetic analogues. Journal of Colloid and Interface Science. 2017;**508**:603-616

[10] Mohamed AMA, Abdullah AM, Younan NA. Corrosion behavior of superhydrophobic surfaces: A review. Arabian Journal of Chemistry. 2015;**8**:749-765

[11] Féat A, Federle W, Kamperman M, van der Gucht J. Coatings preventing insect adhesion: An overview. Progress in Organic Coating. 2019;**134**:349-359

[12] Prianka TR, Subhan N, Reza HM, Hosain MK, Rahman MA, Lee H, et al. Recent exploration of bio-mimetic nanomaterial for potential biomedical applications. Materials Science & Engineering. C. 2018;**93**:1104-1115

[13] Nguyen-Tri P, Tran HN, Plamondon CO, Tuduri L, Vo D-VN, Nanda S, et al. Recent progress in the preparation, properties and applications of superhydrophobic nano-based coatings and surfaces: A review. Progress in Organic Coating. 2019;**132**:235-256

[14] Fihri A, Bovero E, Al-Shahrani A, Al-Ghamdi A, Alabedi G. Recent progress in superhydrophobic coatings used for steel protection: A review. Colloids and Surfaces A: Physicochemical and Engineering Aspects. 2017;**520**:378-390 This paper reported first experimental demonstration of artificial superhydrophobic surfaces at Kao

[15] Syafiq A, Pandey AK, Adzman NN, Abd Rahim N. Advances in approaches and methods for self-cleaning of solar photovoltaic panels. Solar Energy. 2018;**162**:597-619

[16] Selim MS, Higazy SA, Hao Z, Chen X. Chapter 6 - Ceramic polymer nanocomposites as eco-friendly marine antifouling coatings. In: Gupta RK, Nadda AK, Nguyen TA, Sharma S, Bilal M, editors. Micro and Nano Technologies, Advances in Nanotechnology for Marine Antifouling. Amsterdam: Elsevier; 2023. pp. 117-140

[17] Song J, Wang D, Hu L, Huang X, Chen Y. Superhydrophobic surface fabricated by nanosecond laser and perhydropolysilazane. Applied Surface Science. 2018;**455**:771-779

[18] Dimitrakellis P, Gogolides E. Hydrophobic and superhydrophobic surfaces fabricated using atmospheric pressure cold plasma technology: A review. Advances in Colloid and Interface Science. 2018;**254**:1-21

[19] Sun T, Feng L, Gao X, Jiang L. Bioinspired surfaces with special wettability. Accounts of Chemical Research. 2005;**38**(8):644-652

[20] Selim MS, Shenashen MA, Fatthallah NA, Elmarakbi A, El-Safty SA. In situ fabrication of one-dimensional-based lotus-like silicone/γ–Al_2O_3 nanocomposites for marine fouling release coatings. ChemistrySelect. 2017;**2**(30):9691-9700

[21] Jishnu A, Jayan JS, Saritha A, Sethulekshmi AS, Venu G. Superhydrophobic graphene-based materials with self-cleaning and anticorrosion performance: An appraisal of neoteric advancement and future perspectives. Colloids and Surfaces A: Physicochemical and Engineering Aspects. 2020;**606**:125395

[22] Selim MS, El-Safty SA, Azzam AM, Shenashen MA, El-Sockary MA, Abo Elenien OM. Superhydrophobic silicone/TiO_2–SiO_2 Nanorod-like composites for marine fouling release coatings. ChemistrySelect. 2019;**4**:3395

[23] Gao X, Guo Z. Biomimetic Superhydrophobic surfaces with transition metals and their oxides: A review. Journal of Bionic Engineering. 2017;**14**(3):401-439

[24] Singh D, Rehman N, Pandey A. The alternative and efficient solution to biofouling in the aquaculture industry. Applied Biochemistry and Biotechnology. 2023;**195**:4637-4652

[25] Rao TS. Chapter 26 - Biofouling (macro-fouling) in seawater intake systems. In: Water-Formed Deposits, Fundamentals and Mitigation Strategies. Amsterdam: Elsevier; 2022. pp. 565-587

[26] de Carvalho CCR. Marine biofilms: A successful microbial strategy with economic implications. Frontiers in Marine Science. 2018;**5**:1-11

[27] Ombelli M, Costello L, Postle C, Anantharaman V, Meng QC, Composto RJ, et al. Competitive protein adsorption on polysaccharide and hyaluronate modified surfaces. Biofouling. 2011;**27**(5):505-511

[28] Kingsley AB, Muthukrishnan P, Shainy NK, Usha R. Approaches to prevent and treat bacterial biofilm on implantable. Bulletin of Environment, Pharmacology and Life Sciences. 2020;**10**(1):168-174

[29] Yan H, Wu Q, Yu C, Zhao T, Liu M. Recent progress of biomimetic antifouling surfaces in marine. Advanced Materials Interfaces. 2020;**7**(20):1-18

[30] Abarzua S, Jakubowski S. Biotechnological investigation for the prevention of biofouling. I. Biological and biochemical principles for the

prevention of biofouling. Marine Ecology Progress Series. 1995;**123**:301-312

[31] Naser HA. Variability of marine macrofouling assemblages in a marina and a mariculture Centre in Bahrain, Arabian Gulf. Regional Studies in Marine Science. 2017;**16**:162-170

[32] Wang Z, Wang X, Huang Y, Zhou H, Wu Y, Sun Y, et al. Macrofouling organisms: Protection or damage of steel in marine environments? Corrosion Science. 2023;**212**:110928

[33] Pulletikurthi V, Esquivel-Puentes HA, Cheng S, Chamorro LP, Castillo L. Impact of flow regime on the performance of anti-biofouling coatings. Scientific Reports. 2023;**13**(1):1-10

[34] Selim MS, Shenashen MA, El-SockaryMA, FatthallahNA, HigazySA, El-Saeed AM, et al. Chapter 9 - Polymer/graphene-derived nanocomposites as advanced marine antifouling coatings. In: Gupta RK, Nadda AK, Nguyen TA, Sharma S, Bilal M, editors. Micro and Nano Technologies. Advances in Nanotechnology for Marine Antifouling: Elsevier; 2023. pp. 193-230

[35] Davidson I, Cahill P, Hinz A, Kluza D, Scianni C, Georgiades EE. A review of biofouling of ships' internal seawater systems. Frontiers in Marine Science. 2021;**8**:1-16

[36] Almeida VP, Fonseca G. Early biofouling colonization stages: Implications for operation and maintenance planning in marine renewable energy projects. Open Research Europe. 2022;**2**:108

[37] Jiaa R, Unsal T, Xu D, Lekbach Y, Gu T. Microbiologically influenced corrosion and current mitigation strategies: A state-of-the-art review. International Biodeterioration & Biodegradation. 2019;**137**:42-58

[38] Blackwooda DJ, Limb CS, Teo SLM, Xiaoping H, Jianjun P. Macrofouling induced localized corrosion of stainless steel in Singapore seawater. Corrosion Science. 2017;**129**:152-160

[39] Amendola R, Acharjee A. Microbiologically influenced corrosion of copper and its alloys in anaerobic aqueous environments: A review. Frontiers in Microbiology. 2022;**13**:1-10

[40] Selim MS, Shenashen MA, El-Safty SA, Higazy SA, Elmarakbi MA. Review: Progress in biomimetic leverages for marine antifouling using nanocomposite coatings. Journal of Materials Chemistry B. 2020;**8**:3701

[41] Selim MS, El-Safty SA, Shenashen MA. Chapter 8 - Superhydrophobic foul resistant and self-cleaning polymer coating. In: Superhydrophobic Polymer Coatings. Fundamentals, Design, Fabrication, and Applications. Amsterdam: Elsevier; 2019. pp. 181-203

[42] Qureshi AS, Ramaswamy R. Evolution of antifouling paints. International Journal of Science and Research (IJSR). 2017;**6**(10):433-443.

[43] Pei X, Ye Q. Chapter 6: Development of marine antifouling coatings. In: Zhou F, editor. Antifouling Surfaces and Materials. Berlin, Heidelberg: Springer; 2015. pp. 135-149

[44] Hellio C, Yebra D. Advances in Marine Antifouling Coatings and Technologies. Cambridge: Woodshead Publishing; 2009. pp. 1-15

[45] Almeida E, Diamantino TC, de Sousa O. Marine paints: The particular case of antifouling paints. Progress in Organic Coating. 2007;**59**:2-20

[46] Callow ME. Ship fouling: Problems and solutions. Chemistry & Industry. 1990;**5**:123-127

[47] BaSECO. Tributyltin pollution on a global scale. An overview of relevant and recent research: Impacts and issues. In: IMO, Marine Environment Protection Committee. United Kingdom, UK: World Wildlife Fund; 2006

[48] Jin H, Tian L, Bing W, Zhao J, Ren L. Bioinspired marine antifouling coatings: Status, prospects, and future. Progress in Materials Science. 2022;**124**:100889

[49] Liu M, Li S, Wang H, Jiang R, Zhou X. Research progress of environmentally friendly marine antifouling coatings. Polymer Chemistry. 2021;**12**(26):3702-3720

[50] Kyei SK, Asante-Sackey D, Danso-Boateng E. Chapter 8 - Biofouling in the petroleum industry. In: Gupta RK, Nadda AK, Nguyen TA, Sharma S, Bilal M, editors. Micro and Nano Technologies, Advances in Nanotechnology for Marine Antifouling. Amsterdam: Elsevier; 2023. pp. 165-191

[51] Sieminski A. International energy outlook. Energy information administration (EIA). 2014;**18**:2

[52] McCarthy AH, Peck LS, Hughes KA, Aldridge DC. Antarctica: The final frontier for marine biological invasions. Global Change Biology. 2019;**25**:2221-2241

[53] Schultz MP, Bendick JA, Holm ER, Hertel WM. Economic impact of biofouling on a naval surface ship. Biofouling. 2011;**27**(1):87-98

[54] de Oliveira DD, Rojas EG, dos Santos Fernandez MA. Should TBT continue to be considered an issue in dredging port areas? A brief review of the global evidence. Ocean Coastal Management. 2020;**197**:105303

[55] Wei Q, Wu C, Zhang J, Cui Z, Jiang T, Li J. Fabrication of surface microstructure for the ultrafiltration membrane based on "active–passive" synergistic antifouling and its antifouling mechanism of protein. Reactive and Functional Polymers. 2021;**169**:105068

[56] Li Z, Guo Z. Bioinspired surfaces with wettability for antifouling application. Nanoscale. 2019;**11**:22636-22663

[57] Kumar A, Al-Jumaili A, Bazaka O, Ivanova EP, Levchenko I, Bazakaaef K, et al. Functional nanomaterials, synergisms, and biomimicry for environmentally benign marine antifouling technology. Materials Horizons. 2021;**8**:3201-3238

[58] Xie Q, Pan J, Ma C, Zhang G. Dynamic surface antifouling: Mechanism and systems. Soft Matter. 2019;**15**:1087-1107

[59] Sfameni S, Rando G, Marchetta A, Scolaro C, Cappello S, Urzì C, et al. Development of eco-friendly hydrophobic and fouling-release coatings for blue-growth environmental applications: Synthesis, mechanical characterization and biological activity. Gels. 2022;**8**:528

[60] Selim MS, El-Safty SA, El-Sockary MA, Hashem AI, Abo Elenien OM, El-Saeed AM, et al. Smart photo-induced silicone/TiO_2 nanocomposites with dominant [110] exposed surfaces for self-cleaning foul-release coatings of ship hulls. Materials and Design. 2016;**101**:218-225

[61] Li Y, Zhang LY, Zhang C, Zhang Z-R, Liu L. Bioinspired antifouling Fe-based amorphous coating via killing-resisting dual surface modifications. Scientific Reports. 2022;**12**:819

[62] Lagerström M, Wrange A-L, Oliveira DR, Granhag L, Larsson AI, Ytreberg E. Are silicone foul-release coatings a viable and environmentally sustainable alternative to biocidal antifouling coatings in the Baltic Sea region? Marine Pollution Bulletin. 2022;**184**:114102

[63] Carrier AJ, Carve M, Shimeta J, Walker TR, Zhang X, Oakes KD, et al. Transitioning towards environmentally benign marine antifouling coatings. Frontiers in Marine Science. 2023;**10**:1175270

[64] Leonardi AK, Medhi R, Zhang A, Düzen N, Finlay JA, Clarke JL, et al. Investigation of N-substituted morpholine structures in an amphiphilic PDMS-based antifouling and fouling-release coating. Biomacromolecules. 2022;**23**(6):2697-2712

[65] Han X, Wu J, Zhang X, Shi J, Wei J, Yang Y, et al. Special issue on advanced corrosion-resistance materials and emerging applications. The progress on antifouling organic coating: From biocide to biomimetic surface. Journal of Materials Science and Technology. 2021;**61**:46-62

[66] Zhang C, Qi Y, Guo Y, Zhang S, Xiong G, Wang K, et al. Anti-marine biofouling adhesion performance and mechanism of PDMS fouling-release coating containing PS-PEG hydrogel. Marine Pollution Bulletin. 2023;**194**(Part A):115345

[67] Ba M, Zhang Z, Qi Y. Fouling release coatings based on polydimethylsiloxane with the incorporation of phenylmethylsilicone oil. Coatings. 2018;**8**:153

[68] Guazzelli E, Perondi F, Criscitiello F, Pretti C, Oliva M, Casu V, et al. New amphiphilic copolymers for PDMS-based nanocomposite films with long-term marine antifouling performance. Journal of Materials Chemistry B. 2020;**8**:9764-9776

[69] Banerjee I, Pangule RC, Kane RS. Antifouling coatings: Recent developments in the design of surfaces that prevent fouling by proteins, bacteria, and marine organisms. Advanced Materials. 2011;**23**:690-718

[70] Zhou H, Zheng Y, Li M, Ba M, Wang Y. Fouling release coatings based on acrylate–MQ silicone copolymers incorporated with non-reactive Phenylmethylsilicone oil. Polymers. 2021;**13**:3156

[71] Pishbin R, Mohammadi T, Tofighy MA. PVDF ultrafiltration membranes containing copper oxide-charcoal based graphene oxide nanohybrids with enhanced performance and antifouling properties. Process Safety and Environment Protection. 2023;**176**:180-195

[72] Fan F-X, Zheng Y-M, Ba M, Wang Y-F, Kong J-J, Liu J-H, et al. Long time super-hydrophobic fouling release coating with the incorporation of lubricant. Progress in Organic Coating. 2021;**152**:106136

[73] Yang WJ, Neoh K-G, Kang E-T, Teo SL-M, Rittschof D. Polymer brush coatings for combating marine biofouling. Progress in Polymer Science. 2014;**39**(5):1017-1042

[74] Soleimani S, Jannesari A, Etezad SM. Prevention of marine biofouling in the aquaculture industry by a coating based on polydimethylsiloxane-chitosan and sodium polyacrylate. International Journal of Biological Macromolecules. 2023;**245**:125508

[75] Yan Z, Zhou D, Zhang Q, Zhu Y, Wu Z. A critical review on

fouling influence factors and antifouling coatings for heat exchangers of high-salt industrial wastewater. Desalination. 2023;**553**:116504

[76] Kirschner CM, Brennan AB. Bio-inspired antifouling strategies. Annual Review of Materials Research. 2012;**42**:211-229

[77] Mo Y, Xue P, Xing J, Liu H, Wang J, Liu J. Conventional silicone coating improved by novel bioinspired microcavity arrays: Generating stable air bubble arrays for gas-liquid interface antifouling. Colloids and Surfaces A: Physicochemical and Engineering Aspects. 2022;**644**:128887

[78] Tian L, Jin H, Yin Y. Bioinspired antifouling coatings with topographies. In: Advances in Nanotechnology for Marine Antifouling. Amsterdam: Elsevier; 2023. pp. 15-29

[79] Murthy PS, Nanchariah YV, Venkatnarayanan S, Das A, Mohan TVK. Chapter 13 - Antifouling mechanisms in and beyond nature: Leverages in realization of bioinspired biomimetic antifouling coatings. In: Gupta RK, Nadda AK, Nguyen TA, Sharma S, Bilal M, editors. Micro and Nano Technologies, Advances in Nanotechnology for Marine Antifouling. Amsterdam: Elsevier; 2023. pp. 329-362

[80] Angarano M-B, McMahon RF, Hawkins DL, Schetz JA. Exploration of structure-antifouling relationships of capsaicin-like compounds that inhibit zebra mussel (Dreissena polymorpha) macrofouling. Biofouling. 2007;**23**:295-305

[81] Goswami A, Pillai SC, McGranaghan G. Chapter 23 - Nanoscale polymer-based coatings for applications in marine antifouling. In: Thomas S, George JS, editors. Polymer-Based Nanoscale Materials for Surface Coatings. Amsterdam: Elsevier; 2023. pp. 501-546

[82] Chen L, Qian P-Y. Review on molecular mechanisms of antifouling compounds: An update since 2012. Marine Drugs. 2017;**15**:264

[83] Carve M, Scardino A, Shimeta J. Effects of surface texture and interrelated properties on marine biofouling: A systematic review. Biofouling. 2019;**35**:597-617

[84] Bixler GD, Bhushan B. Biofouling: Lessons from nature. Philosophical Transactions of the Royal Society A: Mathematical, Physical and Engineering Sciences. 2012;**370**:2381-2417

[85] Xu Y, Yang L, Zhang H. Bio-inspired surfaces for fouling resistance, a review. E3S Web of Conferences. 2021;**294**:05002

[86] Chung KK, Schumacher JF, Sampson EM, Burne RA, Antonelli PJ, Brennan AB. Impact of engineered surface microtopography on biofilm formation of Staphylococcus aureus. Biointerphases. 2007;**2**:89-94

[87] Chien H-W, Chen X-Y, Tsai W-P, Lee M. Inhibition of biofilm formation by rough shark skin-patterned surfaces. Colloids and Surfaces. B, Biointerfaces. 2020;**186**:110738

[88] Sotoudeh F, Mousavi SM, Karimi N, Lee BJ, Abolfazli-Esfahani J, Manshadi MKD. Natural and synthetic superhydrophobic surfaces: A review of the fundamentals, structures, and applications. Alexandria Engineering Journal. 2023;**68**:587-609

[89] Collins CM, Safiuddin M. Lotus-leaf-inspired biomimetic coatings: Different types, key properties, and applications in infrastructures. Infrastructures. 2022;**7**:46

[90] Selim MS, Shenashen MA, Fatthallah NA, Elmarakbi A, El-Safty SA. In situ fabrication of one-dimensional-based lotus-like silicone/ϒ–Al_2O_3 nanocomposites for marine fouling release coatings. ChemistrySelect. 2017;**2**:9691-9700

[91] Tian L, Yin Y, Jin H, Bing W, Jin E, Zhao J, et al. Novel marine antifouling coatings inspired by corals. Materials Today Chemistry. 2020;**17**:100294

[92] Tian L, Jin E, Yu B, Sun H, Shang Y, Bing W. Novel anti-fouling strategies of live and dead soft corals (*Sarcophyton trocheliophorum*): Combined physical and chemical mechanisms. Journal of Bionic Engineering. 2020;**17**:677-685

[93] Bhushan B, Jung YC. Natural and biomimetic artificial surfaces for superhydrophobicity, self-cleaning, low adhesion, and drag reduction. Progress in Materials Science. 2011;**56**:1-108

[94] Das S, Kumar S, Samal SK, Mohanty S, Nayak SK. A review on superhydrophobic polymer nanocoatings: Recent development and applications. Industrial and Engineering Chemistry Research. 2018;**57**:2727-2745

[95] Rasitha TP, Sofia S, Anandkumar B, Philip J. Long term antifouling performance of superhydrophobic surfaces in seawater environment: Effect of substrate material, hierarchical surface feature and surface chemistry. Colloids and Surfaces A: Physicochemical and Engineering Aspects. 2022;**647**:129194

[96] Laad M, Ghule B. Fabrication techniques of superhydrophobic coatings: A comprehensive review. 2022;**219**(16):2200109

[97] Shi S, Meng S, Zhao P, Xiao G, Yuan Y, Wang H, et al. Underwater adhesion and curing of superhydrophobic coatings for facile antifouling applications in seawater. Composites Communications. 2023;**38**:101511

[98] Bing W, Wang H, Tian L, Zhao J, Jin H, Du W, et al. Small structure, large effect: Functional surfaces inspired by salvinia leaves. Small Structures. 2021;**2**:2100079

[99] Li X, Liu H, Zhang Y, Liu Z. Durable, self-cleaning and anti-fouling superhydrophobic coating based on double epoxy layer. Materials Research Express. 2022;**9**:026404

[100] Selim MS, Elmarakbi A, Azzam AM, Shenashen MA, EL-Saeed AM, El-Safty SA. Eco-friendly design of superhydrophobic nano-magnetite/silicone composites for marine foul-release paints. Progress in Organic Coating. 2018;**116**:21-34

[101] Selim MS, Azzam AM, Higazy SA, El-Safty SA, Shenashen MA. Novel graphene-based ternary nanocomposite coatings as ecofriendly antifouling brush surfaces. Progress in Organic Coating. 2022;**167**:106803

[102] Jeevahan J, Chandrasekaran M, Britto Joseph G, Durairaj R, Mageshwaran G. Superhydrophobic surfaces: A review on fundamentals, applications, and challenges. Journal of Coating Technology and Research. 2018;**15**:231-250

[103] Bae S, Lee YJ, Kim MK, Kwak Y, Choi CH, Kim DG. Antifouling effects of Superhydrophobic coating on sessile marine invertebrates. International Journal of Environmental Research and Public Health. 2022;**19**:7973

[104] Jin H, Wang J, Tian L, Gao M, Zhao J, Ren L. Recent advances in emerging integrated antifouling and

anticorrosion coatings. Materials and Design. 2022;**213**:110307

[105] Li Z, Liu P, Chen S, Liu X, Yu Y, Li T, et al. Bioinspired marine antifouling coatings: Antifouling mechanisms, design strategies and application feasibility studies. European Polymer Journal. 2023;**190**:111997

[106] Si Y, Guo Z. Superhydrophobic nanocoatings: From materials to fabrications and to applications. Nanoscale. 2015;**7**:5922-5946

[107] Nosonovsky M, Bhushan B. Multiscale Dissipative Mechanisms and Hierarchical Surfaces: Friction, Superhydrophobicity, and Biomimetics. Berlin Heidelberg, Germany: Springer-Verlag; 2008

[108] Xiao Y, Zheng J, He Y, Wang L. Droplet and bubble wetting behaviors: The roles of surface wettability and roughness. Colloids and Surfaces A: Physicochemical and Engineering Aspects. 2022;**653**:130008

[109] Young T III. An essay on the cohesion of fluids. Philosophical Transactions of the Royal Society of London; 1805;**95**:65-87

[110] Marmur A, Della Volpe C, Siboni S, Amirfazli A, Drelich JW. Contact angles and wettability: Towards common and accurate terminology. Surface Innovations. 2017;**5**:3-8

[111] Singh MK, Singh A. Chapter 16 - Contact angle and surface wettability measurement. In: Singh MK, Singh A, editors. The Textile Institute Book Series, Characterization of Polymers and Fibres. UK: Woodhead Publishing; 2022. pp. 359-385

[112] Erbil HY, Cansoy CE. Range of applicability of the Wenzel and Cassie-Baxter equations for superhydrophobic surfaces. Langmuir. 2009;**25**:14135-14145

[113] Koishi T, Yasuoka K, Fujikawa S, Ebisuzaki T, Zeng XC. Coexistence and transition between Cassie and Wenzel state on pillared hydrophobic surface. PNAS. 2009;**106**:8435-8440

[114] Ghaffari S, Aliofkhazraei M, Darband GB, Zakeri A, Ahmadi E. Review of superoleophobic surfaces: Evaluation, fabrication methods, and industrial applications. Surfaces and Interfaces. 2019;**17**:100340

[115] Choi W, Tuteja A, Mabry JM, Cohen RE, McKinley GH. A modified Cassie–Baxter relationship to explain contact angle hysteresis and anisotropy on non-wetting textured surfaces. Journal of Colloid and Interface Science. 2009;**339**(1):208-216

[116] Chen L, Duan Y, Cui M, Huang R, Su R, Qi W, et al. Biomimetic surface coatings for marine antifouling: Natural antifoulants, synthetic polymers and surface microtopography. The Science of the Total Environment. 2021;**766**:144469

[117] Damodaran VB, Murthy NS. Bio-inspired strategies for designing antifouling biomaterials. Biomaterials Research. 2016;**20**:18

[118] Asha AB, Ounkaew A, Peng Y-Y, Gholipour MR, Ishihara K, Liu Y, et al. Bioinspired antifouling and antibacterial polymer coating with intrinsic self-healing property. Biomaterials Science. 2023;**11**:128-139

[119] Wang CY, Wang KL, Qian PY, Xu Y, Chen M, Zheng JJ, et al. Antifouling phenyl ethers and other compounds from the invertebrates and their symbiotic fungi collected from the South China Sea. AMB Express. 2016;**6**:1-10

[120] Melrose J. Mucin-like glycopolymer gels in electrosensory tissues generate cues which direct electrolocation in amphibians and neuronal activation in mammals. Neural Regeneration Research. 2019;**14**:1191

[121] Wang P, Zhang D, Lu Z. Slippery liquid-infused porous surface bio-inspired by pitcher plant for marine anti-biofouling application. Colloids and Surfaces, B: Biointerfaces. 2015;**136**:240-247

[122] Xiao L, Li J, Mieszkin S, Di Fino A, Clare AS, Callow ME, et al. Slippery liquid-infused porous surfaces showing marine antibiofouling properties. ACS Applied Materials & Interfaces. 2013;**5**:10074-10080

[123] Masaki T, Fujita D, Hagen N. The surface ultrastructure and epithallium shedding of crustose coralline algae in an 'Isoyake' area of southwestern Hokkaido, Japan. In: Eleventh International Seaweed Symposium. Germany: Springer Nature; 1984. pp. 218-223

[124] Bandyopadhyay PR, Hellum AM. Modeling how shark and dolphin skin patterns control transitional wall-turbulence vorticity patterns using spatiotemporal phase reset mechanisms. Scientific Reports. 2014;**4**:6650

[125] Li J, Guo Z, Liu W. Biomimetic superhydrophobic materials construct from binary structure: A Review on design, properties, and applications. 2023;**10**(2):2201847

[126] Zheng L, Sundaram HS, Wei Z, Li C, Yuan Z. Applications of zwitterionic polymers. Reactive and Functional Polymers. 2017;**118**:51-61

[127] Arabaghian H, Wang M, Ordonez J, Rodrigues DF. Chapter 2 - Zwitterionic polymers in biofouling and inorganic fouling mechanisms. In: Tseng H-H, Lau WJ, Al-Ghouti MA, An L, editors. 60 Years of the Loeb-Sourirajan Membrane. Amsterdam: Elsevier; 2022. pp. 33-70

[128] Nurioglu AG, Esteves ACC. Non-toxic, non-biocide-release antifouling coatings based on molecular structure design for marine applications. Journal of Materials Chemistry B. 2015;**3**:6547-6570

[129] Satdive A, Tayde S, Toksha B, Kundu D, Naik J, Hazra C, et al. Superhydrophobic hybrid nanocomposites: Mapping the current research trends and recent advances. Chemical Engineering Science. 2023;**278**:118941

[130] Selim MS, El-Safty SA, El-Sockary MA, Hashem AI, Abo Elenien OM, EL-Saeed AM, et al. Tailored design of Cu_2O nanocube/silicone composites as efficient foul-release coatings. RSC Advances. 2015;**5**(26):19933-19943

[131] Selim MS, El-Safty SA, El-Sockary MA, Hashem AI, Abo Elenien OM, EL-Saeed AM, et al. Modeling of spherical silver nanoparticles in silicone-based nanocomposites for marine antifouling. RSC Advances. 2015;**5**:63175-63185

[132] Selim MS, Yang H, El-Safty SA, Fatthallah NA, Shenashen MA, Wang FQ, et al. Superhydrophobic coating of silicone/β–MnO_2 nanorod composite for marine antifouling. Colloids and Surfaces, A: Physicochemical and Engineering Aspects. 2019;**570**:518-530

[133] Selim MS, El-Safty SA, Fatthallah NA, Shenashen MA. Silicone/graphene oxide sheet-alumina nanorod ternary composite for superhydrophobic antifouling coating. Progress in Organic Coating. 2018;**121**:160-172

[134] Zhang Z, Chen R, Song D, Yu J, Sun G, Liu Q, et al. Guanidine-functionalized graphene to improve the antifouling performance of boron acrylate polymer. Progress in Organic Coating. 2021;**159**:106396

[135] Pan J, Peng Q, Zhang G, Xie Q, Gong X, Qian P-Y, et al. Antifouling mechanism of natural product-based coatings investigated by digital holographic microscopy. Journal of Materials Science and Technology. 2021;**84**:200-207

[136] Bai XQ, Xie GT, Fan H, Peng ZX, Yuan CQ, Yan XP. Study on biomimetic preparation of shell surface microstructure for ship antifouling. Wear. 2013;**306**:285-295

[137] Ijaola AO, Bamidele EA, Akisin CJ, Bello IT, Oyatobo AT, Abdulkareem A, et al. Wettability transition for laser textured surfaces: A comprehensive review. Surfaces and Interfaces. 2020;**21**:100802

[138] Liu YH, Li GJ. A new method for producing “lotus effect” on a biomimetic shark skin. Journal of Colloid and Interface Science. 2012;**388**:235-242

[139] Kumar SA, Sinhamahapatra A. Chapter 9 - Biomimetics in smart coatings. In: Kumar A, Behera A, Nguyen TA, Bilal M, Gupta RK, editors. Antiviral and Antimicrobial Smart Coatings. Amsterdam: Elsevier; 2023. pp. 263-285

[140] Decker JT, Sheats JT, Brennan AB. Engineered antifouling microtopographies: Surface pattern effects on cell distribution. Langmuir. 2014;**30**:15212-15218

[141] Guan Y, Chen R, Sun G, Liu Q, Liu J, Yu J, et al. Ficus religiosa-inspired microstructure-controlled low surface energy coatings with long-term antifouling effect. Colloids and Surfaces A: Physicochemical and Engineering Aspects. 2023;**670**:131482

[142] Schumacher JF, Carman ML, Estes TG, Feinberg AW, Wilson LH, Callow ME, et al. Engineered antifouling microtopographies - effect of feature size, geometry, and roughness on settlement of zoospores of the green alga Ulva. Biofouling. 2007;**23**:55-62

[143] Mehta H, Kaur G, Chaudhary GR. Environmentally benign fouling release metallosurfactant coatings; evaluation of surface properties, leaching rate, biocorrosion resistance and durability. Applied Surface Science. 2023;**635**:157726

[144] Cao X, Pettitt ME, Wode F, Arpa Sancet MP, Fu J, Ji J, et al. Interaction of zoospores of the green alga Ulva with bioinspired micro- and nanostructured surfaces prepared by polyelectrolyte layer-by-layer self-assembly. Advanced Functional Materials. 2010;**20**:1984-1993

[145] Scardino AJ, de Nys R. Mini review: Biomimetic models and bioinspired surfaces for fouling control. Biofouling. 2011;**27**:73-86

[146] Richards C, Slaimi A, O’Connor NE, Barrett A, Kwiatkowska S, Regan F. Bioinspired surface texture modification as a viable feature of future aquatic antifouling strategies: A review. International Journal of Molecular Sciences. 2020;**21**(14):5063

[147] Zhao L, Chen R, Lou L, Jing X, Liu Q, Liu J, et al. Layer-by-layer-assembled antifouling films with surface microtopography inspired by Laminaria japonica. Applied Surface Science. 2020;**511**:145564

[148] Chen ZF, Zhao WJ, Xu JH, Mo MT, Peng SS, Zeng ZX, et al. Designing

environmentally benign modified silica resin coatings with biomimetic textures for antibiofouling. RSC Advances. 2015;**5**:36874-36881

[149] Qu M, He J, Zhang J. Superhydrophobicity, learn from the lotus leaf. In: Biomimetics Learning from Nature. UK: IntechOpen; 2010. pp. 325-342

[150] Yu H, Zhang J, Zhang S, Han Z. Bionic structures and materials inspired by plant leaves: A comprehensive review for innovative problem-solving. Progress in Materials Science. 2023;**139**:101181

[151] Barthlott W, Neinhuis C. Purity of the sacred lotus, or escape from contamination in biological surfaces. Planta. 1997;**202**:1-8

[152] Yan YY, Gao N, Barthlott W. Mimicking natural superhydrophobic surfaces and grasping the wetting process: A review on recent progress in preparing superhydrophobic surfaces. Advances in Colloid and Interface Science. 2011;**169**(2):80-105

[153] Zhang M, Feng S, Wang L, Zheng Y. Lotus effect in wetting and self-cleaning. Biotribology. 2016;**5**:31-43

[154] Hsu S-H, Woan K, Sigmund W. Biologically inspired hairy structures for superhydrophobicity. Materials Science & Engineering R: Reports. 2011;**72**(10):189-201

[155] Feng L, Li S, Li Y, Li H, Zhang L, Zhai J, et al. Super-hydrophobic surfaces: From natural to artificial. Advanced Materials. 2002;**14**:1857-1860

[156] Barthlott W, Mail M, Neinhuis C. Superhydrophobic hierarchically structured surfaces in biology: Evolution, structural principles and biomimetic applications. Philosophical Transactions of the Royal Society of London, Series A. 2016;**374**:20160191

[157] Celia E, Darmanin T, de Givenchy ET, Amigoni S, Guittard F. Recent advances in designing superhydrophobic surfaces. Journal of Colloid and Interface Science. 2013;**402**:1-18

[158] SunJ, BhushanB. Nanomanufacturing of bioinspired surfaces. Tribology International. 2019;**129**:67-74

[159] Yan C, Jiang P, Jia X, Wang X. 3D printing of bioinspired textured surfaces with superamphiphobicity. Nanoscale. 2020;**12**:2924-2938

[160] Farzin H. Micromachining: A new trend in manufacturing. In: 120th ASEE Annual Conference & Exposition, 23-26 June 2013. Atlanta, USA: American Society for Enginnering Education; 2013. pp. 23.907.1

[161] Subramani K, Ahmed W. Self-assembly of proteins and peptides and their applications in bionanotechnology and dentistry. In: Emerging Nanotechnologies in Dentistry. Second ed. New York, USA: William Andrew Publishing, Applied Science Publishers; 2018. pp. 231-249

[162] Duncan AC, Weisbuch F, Rouais F, Lazare S, Baquey C. Laser microfabricated model surfaces for controlled cell growth. Biosensors & Bioelectronics. 2002;**17**:413-426

[163] Varghese RJ, Parani S, Thomas S, Oluwafemi OS, Wu J. Introduction to nanomaterials: Synthesis and applications. In: Nanomaterials for Solar Cell Applications. Amsterdam: Elsevier; 2019. pp. 75-95

[164] Bixler GD, Bhushan B. Bioinspired rice leaf and butterfly wing surface

structures combining shark skin and lotus effects. Soft Matter. 2012;**8**:11271-11284

[165] Selim MS, Yang H, Wang FQ, Fatthallah NA, Li X, Li Y, et al. Superhydrophobic silicone/SiC nanowire composite as a fouling release coating material. Journal of Coating Technology and Research. 2019;**16**:1165-1180

[166] Selim MS, Yang H, Wang FQ, Fatthallah NA, Huang Y, Kuga S. Silicone/ZnO nanorod composite coating as a marine antifouling surface. Applied Surface Science. 2019;**466**:40-50

[167] Kapridaki C, Maravelaki-Kalaitzaki P. TiO_2–SiO_2–PDMS nano-composite hydrophobic coating with self-cleaning properties for marble protection. Progress in Organic Coating. 2013;**76**(2-3):400

[168] Chen Y, Ao J, Zhang J, Gao J, Hao L, Jiang R, et al. Bioinspired superhydrophobic surfaces, inhibiting or promoting microbial contamination? Materials Today. 2023;**67**:468-494. DOI: 10.1016/j.mattod.2023.06.006

[169] Selim MS, Fatthallah NA, Higazy SA, Hao Z, Mo PJ. A comparative study between two novel silicone/graphene-based nanostructured surfaces for maritime antifouling. Journal of Colloid and Interface Science. 2022;**606**(Part 1):367-383

[170] Soleimani S, Jannesari A, Yousefzadi M, Ghaderi A, Shahdadi A. Eco-friendly foul release coatings based on a novel reduced graphene oxide/Ag nanocomposite prepared by a green synthesis approach. Progress in Organic Coating. 2021;**151**:106107

[171] Selim MS, Fatthallah NA, Shenashen MA, Higazy SA, Madian HR, Selim MM, et al. Bioinspired graphene oxide–magnetite nanocomposite coatings as protective superhydrophobic antifouling surfaces. Langmuir. 2023;**39**(6):2333-2346

Chapter 6

Hydrophobic Protection for Building Materials

Katarzyna Buczkowska

Abstract

The chapter "Hydrophobic Protection for Building Materials" highlights the significance of modifying wetting properties to enhance the durability and performance of construction materials. It addresses the theme of wetting in building materials, wetting assessment techniques, and factors influencing wetting, such as chemical composition, pore distribution, and surface properties. Traditional building materials are compared with innovative materials like geopolymers. Various methods of wetting modification are discussed in this chapter, including altering material composition through bulk additives and coatings. Research is presented to acquaint the reader with current trends in modifying the wetting of construction materials. The chapter underscores the importance of nanomaterials and bulk additives in altering surface properties and outlines investigations into coatings designed for surface protection. Practical applications of hydrophobic coatings are also demonstrated through examples of different commercial products.

Keywords: surface modification, hydrophobic protection, civil engineering materials, geopolymer composites, ecological materials

1. Introduction

The realm of building materials is one of continuous evolution, driven by the pursuit of enhancing their durability, performance, and functionality in the face of diverse environmental conditions. Among the various challenges that materials encounter, the phenomenon of wetting stands out as a critical determinant of their behavior. The concept of wetting, encompassing the ability of materials to absorb fluids and propagate them along their surfaces, holds profound implications for material integrity and longevity. In the context of construction, where materials are subjected to multifaceted conditions, understanding and modifying wetting behavior emerge as crucial pursuits.

Section 3 of this chapter is dedicated to exploring the intricacies of wetting, shedding light on the factors influencing a material's interaction with fluids and elucidating techniques for assessing and modifying wetting characteristics. This foundational understanding serves as a stepping stone toward enhancing material performance in construction applications. As the behavior and durability of building materials are deeply intertwined with wetting phenomena, comprehensive investigation and

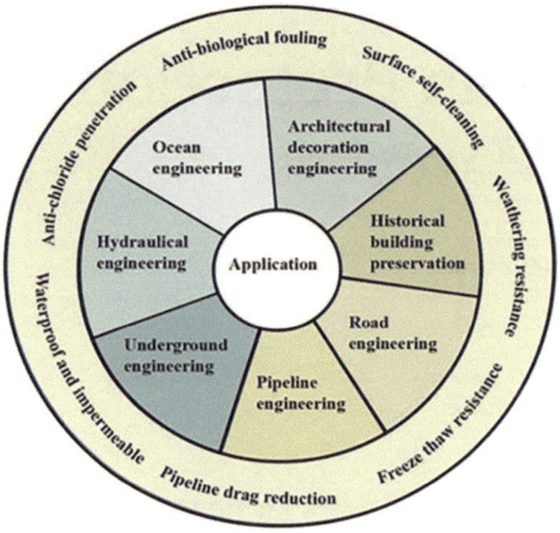

Figure 1.
Applications of hydrophobically modified cement-based materials [1].

manipulation of wetting behaviors become pivotal. **Figure 1** shows the application areas for hydrophobically modified cement-based materials.

The assessment of wetting properties is a pivotal step in deciphering the interaction between building materials and fluids. Contact angle measurement, a key parameter in quantifying wetting behavior, is scrutinized through various techniques. The spreading method, laser beam method, and drop method each offer unique insights into the material's interaction with fluids, unveiling the dynamic interplay between surface and liquid.

The wetting behavior of building materials is governed by a complex interplay of factors. The arrangement of pores within a material's structure holds sway over its ability to absorb fluids, with parameters such as size and distribution significantly affecting the degree of wetting. As revealed by research, the manipulation of pore characteristics can lead to material optimization, ensuring efficient fluid interaction and enhanced durability.

Chemical composition and surface properties further influence wetting behavior. The presence of molecules and functional groups on a material's surface can profoundly alter its interaction with fluids. Through meticulous study and manipulation of these surface characteristics, researchers can tailor material wetting behaviors to suit specific needs, thus fortifying material performance.

In this dynamic landscape, the protection of building materials assumes paramount importance. Techniques such as surface coatings, hydrophobic and oleophobic treatments, and anti-corrosion agents act as guardians, shielding materials from the deleterious effects of external elements. These interventions not only enhance material longevity but also provide avenues for creativity and esthetics in construction.

In summary, the exploration of wetting behaviors in building materials offers a gateway to unraveling the intricate interplay between materials and fluids. This understanding forms the bedrock for fortifying material performance, improving durability, and expanding the horizons of construction applications. Through the careful manipulation of wetting behaviors and the application of protective measures, building materials can transcend their inherent limitations and pave the way for resilient, functional, and visually appealing structures.

2. Comparison of geopolymers and Portland cement

Portland cement is the most commonly used type of cement and a fundamental component in the production of concrete and mortar. It is a finely ground, rocky mixture consisting mainly of calcium oxide (quicklime - 65%) and silica (silica dust - 21%), as well as, for example, alumina (5.6%) or iron oxide (3.4%). These substances occur primarily in the following forms [2]:

- Tricalcium silicate - $(CaO)_3$. SiO_2, commonly referred to as alite, often denoted as C_3S.
- Dicalcium silicate - $(CaO)_2$. SiO_2, commonly referred to as belite, often denoted as C_2S.
- Tetra-calcium aluminoferrite - $(CaO)_4$. Al_2O_3. Fe_2O_3, commonly referred to as celite, often denoted as C_4AF.
- Tricalcium aluminate - $(CaO)_3$. Al_2O_3, commonly denoted as C_3A.

It is produced by grinding and burning a mixture of calcium carbonate (in the form of limestone, chalk, or other rocks), silica (in the form of sand or clay, but old glass can also be used), bauxite (as a source of alumina, but recycled aluminum can also be used), iron ore (as a source of iron oxide, but recycled iron or fly ash from power plants can also be used), gypsum, and other additives. This mixture is hydrated after dissolution in water and solidifies by forming calcium hydroxide (allowing the cement to set even underwater). Therefore, the main difference between geopolymers and Portland cement lies in the reaction that occurs during solidification. Geopolymers undergo polycondensation, while cement undergoes hydration. Cement can simply be mixed with water, whereas geopolymers require activating solutions to achieve a high pH. These activating solutions increase the cost of geopolymers [3, 4].

Portland cement is the primary ingredient in concrete (specifically known as cement concrete - alternatively, asphalt, gypsum, or other substances can be used instead of cement). Concrete is the most widely used composite material in construction worldwide. Hydrated cement serves as the binder and matrix. Additionally, concrete contains aggregates (often called coarse aggregates, used to create a durable compressive skeleton formed by interlocking individual grains, thereby increasing the overall compressive strength). Commonly used aggregates include sand, gravel, or crushed stone with various particle sizes. Other additives such as metal or polymer fibers (used to increase tensile and flexural strength), other mineral or liquid admixtures, and accelerators or superplasticizers (to improve the workability and consistency of the concrete mix) are also added. The combination of concrete and iron or steel reinforcement allows for reinforced concrete, which can better withstand tensile loads than plain concrete [5, 6].

Geopolymers have significantly better mechanical properties compared to concrete made from Portland cement, particularly much higher compressive strength (around 100 MPa for geopolymers, 30 MPa for cement, and up to 60 MPa for cement with special modifications). However, tensile and flexural strength are lower in geopolymers. This can be compensated for by using reinforcements commonly used in conventional concrete, such as carbon steel bars or corrosion-resistant materials like carbon fiber. However, it is not possible to use any lightweight metals or alloys in the production of geopolymer composites as these materials quickly lose their strength due to the strong alkalinity of geopolymers. To produce geopolymer composites with glass fibers, it is necessary to use alkali-resistant glass fibers, for example, made of organic polymers [4, 7].

One of the most significant advantages of geopolymers is their higher thermal resistance. While concrete rapidly degrades when heated to temperatures above 300° C due to thermal decomposition, which also releases toxic substances, geopolymers remain stable up to a melting temperature of around 1265°C. However, even in this case, some changes occur in the structure or composition. In the temperature range of 100–200°C, there is an increase in porosity and a decrease in mass due to the release of bound water. As the temperature increases, there is further mass loss. At temperatures above 600°C, siloxane bonds begin to break down, releasing hydroxyl ions. With further temperature increase (above 800°C), porosity decreases due to partial melting or sintering of geopolymers into a more compact structure. However, these changes are much less destructive than the changes in concrete exposed to too high temperatures [4, 7].

Geopolymers also have low thermal conductivity, which can be further reduced by foaming. This can be achieved by adding powdered aluminum, which reacts with alkaline substances, such as hydroxides in the liquid geopolymer mixture, releasing hydrogen gas. All these properties, therefore, make it feasible to use geopolymers for passive fire protection. [4, 7].

Geopolymers are also much more resistant to chemical attack, making them suitable for use in harsh conditions, such as acidic rain, sewage treatment plants, or chimneys. Moreover, they exhibit significant adsorptive capabilities, making them potentially useful, for example, in wastewater treatment, where they can serve as adsorbents or filtering media. Additionally, they can seal various particles incorporated into the structure during geopolymerization, thus avoiding leakage into the surrounding environment. Consequently, they can be used for storing hazardous waste or other waste materials, such as exhaust gases from industrial plants or power plants [4, 7].

However, the main advantage of geopolymers lies in their much lower energy consumption during production and lower emissions. The production of geopolymer cement requires approximately 1230–1310 MJ/t, with most of the energy consumed during the calcination of clay materials. In contrast, the production of Portland cement consumes nearly three times more energy, about 3500 MJ/t, especially during the calcination of limestone. Furthermore, Portland cement production generates a large amount of carbon dioxide (approximately one ton per ton of cement), which cannot be significantly reduced since carbon dioxide is an unavoidable byproduct of limestone calcination. In 2005, cement production was responsible for 1.8 billion tons of carbon dioxide emissions, accounting for about 8% of the total emissions that year. In contrast, geopolymer production emits 50–80% less carbon dioxide, mainly due to lower temperature requirements (which reduces energy demand). Temperatures up to 1500°C are required for cement production, while 600–700°C are sufficient for thermal processing of clay components in geopolymers [8].

The main drawback of geopolymers is the need for their alkaline activation, which poses a logistical problem. This is due to the necessity of purchasing or producing, distributing, and more complex on-site use of activating solutions because of their strong alkalinity (corrosiveness). Materials based on ordinary Portland cement simply need to be mixed with water. Another disadvantage is the risk of variability in the composition and structure of materials used in geopolymer production, especially fly ash or slag. The variable composition of activators and binders leads to inconsistent properties of the obtained material, which is problematic in standardizing these materials, whereas the composition and properties of Portland cement are clearly defined. However, this problem can be addressed.

3. Wettability of building materials

The concept of wettability constitutes a fundamental aspect that profoundly influences the behavior and durability of construction materials, especially when exposed to various environmental conditions. This dedicated section of the chapter aims to delve into the nuances of wettability, examining numerous factors that affect a material's ability to absorb fluids. Furthermore, it elucidates available techniques for assessing and modifying wettability characteristics. In essence, wettability pertains to the inherent tendency of a surface to absorb and facilitate the spread of fluids along its structure. Unlike droplet formation, this phenomenon involves fluids like water adhering to the material's surface [1, 9].

The fundamental tool for expressing the level of wettability is the contact angle. This angle measures the inclination of a liquid droplet on the material's surface. A smaller contact angle indicates strong wettability, where the fluid spreads widely. This implies a high surface energy. On the other hand, a larger contact angle signifies weak wettability, where the fluid tends to form droplets and roll off the surface due to lower surface energy. Through this measure, researchers can discern a material's propensity for fluid absorption, its capillary action potential, or repellency, and how these properties directly impact its durability and overall performance [9, 10].

Hao Yao and colleagues [9], in their publication, focused on comprehending the mechanisms of wettability transition on microstructural surfaces, revealing the intricate interdependence between surface geometry and wettability behavior. By combining

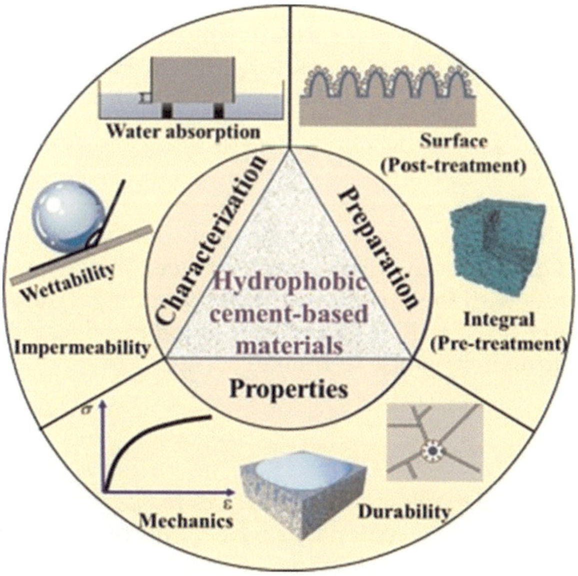

Figure 2.
Illustration of hydrophobic cementitious materials from three aspects, covering preparation, characterization, and properties [9].

theoretical analysis with experimental methods, researchers illustrate how microstructures can be designed to induce specific wettability effects, potentially revolutionizing material design. **Figure 2** presents an illustration of hydrophobic cementitious materials from three aspects: preparation, characterization, and properties.

The composition and microstructural organization of the material have an impact on wettability behavior. Jihui Zhao et al. [1] investigate the effects of hydrophobic modifications on concrete and their influence on surface free-energy. François-Xavier Coudert and colleagues [11] analyze changes in wettability properties and surface free energy due to concrete hydrophobization. Meanwhile, Chenzhi Li and colleagues [12] focus on the mechanisms of wettability transition on microstructural surfaces and their significance for the durability of construction materials.

The influence of pore structure on wettability has been described by Jihui Zhao and his team [13], emphasizing the importance of surface analysis techniques such as X-ray photoelectron spectroscopy (XPS) and surface energy analysis (SEA) in understanding and manipulating wettability. Similarly, Wang et al. [14] concentrate on testing and modeling wettability transition on microstructural surfaces.

In summary, wettability serves as a foundation for understanding the behavior and durability of building materials. The mentioned publications highlight the multidisciplinary nature of wettability's influence, encompassing microstructural considerations, surface modifications, and broader material applications. Together, these studies contribute to a more precise analysis of the impact of wettability on material behavior and their diverse applications in the field of construction.

3.1 Methods for assessing wettability

The assessment of wetting in building materials is a crucial step in understanding how fluids interact with their surfaces. This assessment is based on the measurement of the contact angle – the angle at which a liquid droplet rests on the material's surface. This angle provides insights into the wetting behavior of the material, offering information about the spreading or droplet formation of fluids on the surface. Various techniques for measuring the contact angle are utilized, each revealing different aspects of the material's interaction with fluids. These techniques include: [15, 16].

3.1.1 Spreading method

In this method, the change in contact angle over time is carefully monitored by introducing a liquid droplet onto the material's surface. This dynamic analysis reveals how the fluid interacts with the surface over time, shedding light on the material's ability to absorb or repel fluids. The spreading method provides subtle insights into the evolution of wetting in response to fluid exposure.

3.1.2 Laser beam method

The laser beam method employs a laser beam to measure the angle of inclination of a liquid droplet resting on the material's surface. This technique offers precise measurement of the contact angle, allowing for an accurate assessment of wetting characteristics. The precision of the laser beam ensures reliable quantification of contact angles, contributing to a thorough understanding of the material's response to fluids.

3.1.3 Drop method

The drop method takes a different approach by capturing micrographs of droplets on the material's surface. The shape and dimensions of these droplets provide significant information about the material's wetting tendencies. Analyzing the geometry of droplets – whether they spread or maintain their shape – offers a qualitative and visual representation of how fluids interact with the surface.

Shirong Zhu et al. [16] conducted extensive research on advanced methods for assessing surface-free energy and wetting. Their work delves into the intricacies of these measurement techniques, offering insights into their applications, advantages, and limitations. This comprehensive review serves as a valuable resource for researchers and practitioners seeking to understand the complexity of wetting measurement.

Fan Ding and Manglai Gao [17] examined wetting measurement methods to enhance oil recovery. Although focused on a different context, their review provides valuable insights into various techniques used across different fields for wetting assessment. This broad perspective highlights the versatility of wetting measurement and its applications beyond the realm of building materials.

Significant research was conducted by Grzegorczyk-Frańczak et al. [18]. Their work focused on the changes in wetting properties and surface energy during the hydrophobization of concretes with the addition of boiler slag and coal dust. These studies provided valuable insights into the influence of these additives on the hydrophobic properties of concrete, which is crucial for the durability and performance of building materials under various environmental conditions.

On the other hand, the work of Adhikary et al. [19] provided a review on lightweight self-compacting concrete. The authors concentrated on analyzing various aspects of this type of concrete, including its components, properties, and potential applications in construction. Their research emphasizes the importance of lightweight self-compacting concrete in the context of sustainable construction and the need for further studies to optimize its properties and applications.

In conclusion, wetting measurement through contact angle assessment forms the basis for uncovering the intricate dynamics between materials and fluids. The presented methods showcase a multifaceted approach to understanding wetting behavior. These measurement techniques, supported by in-depth research, serve as essential tools for delving into the interaction of building materials with fluids, ultimately influencing their performance and durability in construction applications.

3.2 Factors influencing wettability

The wetting of building materials is a multifaceted phenomenon shaped by a range of factors related to their structural, chemical, and physical properties. Within this complex structure, several key determinants play vital roles in shaping the wetting characteristics of the material, each offering a unique perspective on its behavior during interactions with fluids. An essential aspect of these properties is wettability, which refers to the material's capacity to absorb liquids, such as water. The influence of the material's structure on its ability to absorb fluids has been the subject of numerous studies and analyses, contributing to a better understanding of this phenomenon.

3.2.1 Pores - size and distribution

The significance of pore distribution and size in building materials regarding their interaction with fluids, particularly water, is well-documented in scientific literature. **Figure 3** shows the complexity of the correlation study of concrete pore structure based on WOS (2018–2023). The distribution of pores within the material has a significant impact on various aspects of its interaction with fluids. It is often observed that materials with a higher percentage of larger pores exhibit the ability to absorb a greater amount of water and other fluids. However, the quantity of pores is not the only crucial factor; their size and distribution also play a key role. These parameters determine to what extent fluids can penetrate the material. For example, materials characterized by an organized network of small pores often demonstrate much better wettability compared to materials with a more chaotic pore structure.

In a study conducted by Song Jin et al. [21], the impact of pore structure on the ability of concrete with varying strengths to absorb water and the permeability of chloride was analyzed. Another example comes from the research of Jianzhuang Xiao et al. [20]. Their work focused on investigating the influence of pore structure on concrete's ability to absorb water. These studies revealed that the distribution and distribution of pores have a complex impact on the material's response to varying humidity levels. The results provide valuable insights into optimizing concrete properties related to water interaction. In the studies presented by Dinghua Zou et al. [22], the authors demonstrate that the pore structure of the aggregate affects the equilibrium of residual water, but has a negligible impact on the rate of water release.

Furthermore, it is worth noting that the influence of pore structure on the ability to absorb water is not limited to traditional materials such as concrete. Shikun Chen et al. [23] conducted research on the pore structure and water absorption capability of

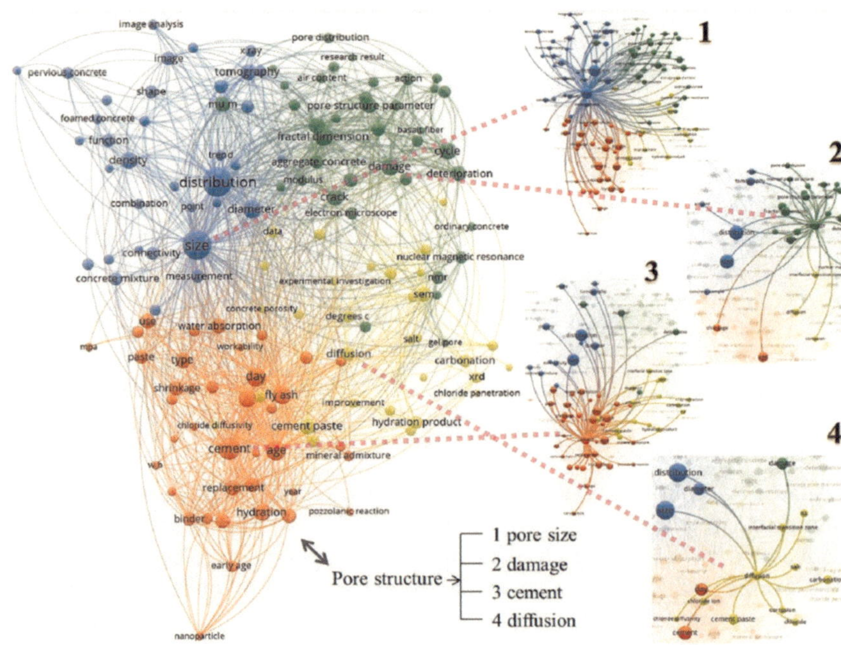

Figure 3.
Correlation of concrete pore structure study based on WOS (2018–2023) [20].

geopolymers. In the context of studying the influence of distribution, size, and pore structure on the hydrophobic properties of alkali-activated materials, it is essential to consider the work of Bang et al. [24]. Another study by Ruan et al. [25] focused on the analysis of the microstructure and mechanical properties of sustainable cementitious materials with an ultra-high substitution level of limestone and calcined clay powder.

The research presented by Lawrence M. et al. [26] offers tangible insights into how pore distribution influences a material's water absorption behavior. This study underscores not only the pivotal role of pore structure in interaction processes but also accentuates the imperative to comprehend and regulate these characteristics to enhance the efficacy of building materials. Grasping these facets is paramount for engineers and scientists striving to develop materials with optimal characteristics concerning water resistance and longevity in diverse environmental scenarios.

The size and distribution of pores in building materials significantly influence their interaction with fluids, especially water. **Figure 3** illustrates the complex relationships between the pore structure in concrete and its interaction with fluids, summarizing publications from WoS between 2018 and 2023. Materials with a higher proportion of larger pores tend to absorb more water. However, the sheer number of pores is not the only determinant; their size and distribution are equally important. Materials with a well-organized network of small pores often exhibit better wettability than those with a random pore structure. Numerous studies have focused on the impact of pore structure on concrete's ability to absorb water and other properties. Notably, the influence of pore structure is not limited to traditional materials like concrete but also extends to newer materials such as geopolymers. Understanding the role of pore distribution and size is crucial for optimizing water resistance and durability of building materials in various environmental conditions.

3.2.2 Chemical composition and surface properties

The wettability of materials is shaped not only by their physical structure but also by their chemical composition (**Figure 4**) and surface properties (**Figure 5**). Molecules and functional groups present on the material's surface can interact with fluids in various ways, directly affecting the material's absorption capacity. This complex interplay between surface chemistry and wetting behavior has been thoroughly investigated and documented in various scientific studies. The aforementioned issue will be discussed in detail in Section 4 of the above chapter.

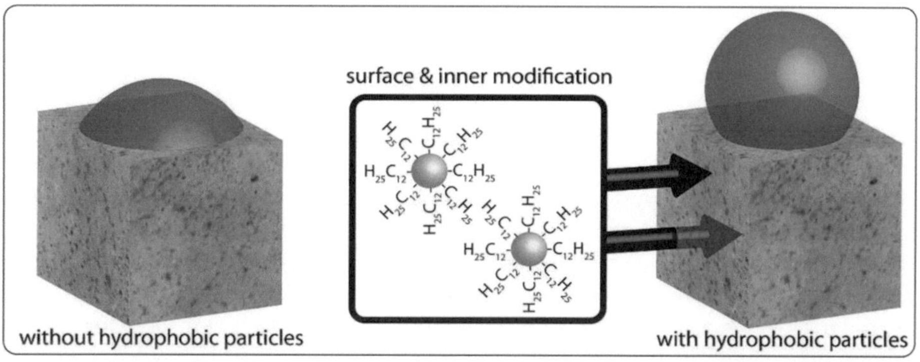

Figure 4.
Effect of chemical composition modification on wetting properties [27].

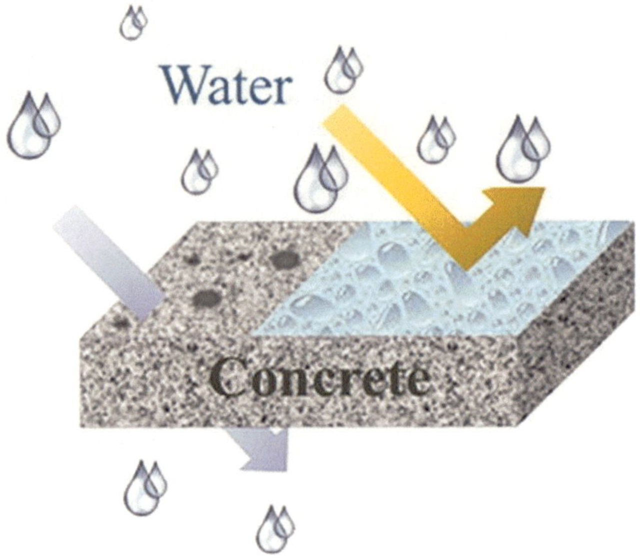

Figure 5.
Effect of surface modification on wetting properties [28].

4. Methods of protecting building materials

In the field of construction, modifying the wettability of materials plays a crucial role in improving their durability, performance, and functionality. The literature presents various approaches that contribute to understanding ways to shape wettability to enhance material performance. One approach involves adding additives to modify the material's structure and improve wettability, while another direction involves applying appropriate hydrophobic coatings.

The following information presents methods of modifying material wettability through both additives and applied coatings. By understanding and utilizing these approaches, researchers and engineers can create materials with tailored surface properties that positively impact their durability and performance across various applications in construction and beyond.

4.1 Bulk additives in modifying wettability of building materials

The way building materials interact with water and their ability to withstand it are of paramount importance to their strength and durability. This section focuses on altering material behavior by modifying its structure through additives. As people seek materials with improved properties, researchers investigate how to modify the surface properties of these materials.

Research of Shidong Li et al. [29] delves deep into the influence of silica nanofluids on wettability alteration. By employing advanced visualization techniques, Li et al. were able to provide a comprehensive understanding of how surface properties, when interacted with specific chemical compositions, can significantly alter a material's ability to absorb water. Their findings underscore the importance of nano-level interactions and their profound impact on macro-level properties.

In the multifaceted exploration of water absorption in materials, the incorporation of recycled materials into building mixes offers a unique dimension. A pivotal study in this context was conducted by Alena Sicakova et al. [30]. They examined the characteristics of building mixes that incorporate fine particles from selected recycled materials. Sicakova and her team's findings suggest that the inclusion of these recycled particles can significantly influence the water absorption properties of the resulting composite. This highlights not only the potential for sustainable material development but also emphasizes the intricate interplay between chemical composition and surface properties when recycled components are introduced. Their research lays the groundwork for understanding the potential benefits and challenges of integrating recycled materials into building mixes, offering a fresh perspective on optimizing water absorption properties while championing sustainability.

A seminal study by Ernesto Mora et al. [27] provides invaluable insights into this endeavor. Their research delves into the control of water absorption in concrete materials through modification with hybrid hydrophobic silica particles. The introduction of these silica particles alters not only the inherent properties of the concrete but also significantly reduces its water absorption capacity. Mora and his team's findings underscore the transformative potential of such modifications, suggesting a promising avenue for enhancing the durability and longevity of construction materials.

In summary, wettability, a key property of materials, is shaped by both physical structure and a complex balance of chemical composition and surface properties. The behavior of building materials in contact with liquids results from a complex interaction of factors, each contributing a unique aspect to the material's reaction to interactions with liquids - pore structure, surface chemical composition, and type of surface. All these factors together form the wettability profile of the material. This subtle interdependence is fundamental to understanding how building materials interact with liquids, which has profound implications for their durability and performance in various construction applications. Molecules and functional groups on the material's surface play a crucial role in determining how it interacts with liquids (**Figure 4**). Recent research sheds light on this intricate relationship. Shidong Li and his team [29] highlighted the role of silica nanofluids in altering wettability, emphasizing the importance of interactions at the nano level. Meanwhile, studies conducted by Sicakova et al. [30] revealed the potential of incorporating recycled materials into construction mixes, showing how such additives can affect water absorption properties and promote sustainable development. Additionally, the works of Mory, González, Romero, and Castellón [27] demonstrated the potential of modifying concrete with hydrophobic silica particles to reduce its water absorption. These findings underscore the importance of understanding and manipulating surface chemistry to design materials with tailored wettability properties, suitable for a wide range of applications in construction and beyond.

4.2 Surface protection of building materials

Surface protection of building materials is a crucial element that ensures the durability and longevity of construction. Building materials such as concrete, sandstone, bricks, and ceramics are exposed to various external factors, including water, moisture, salts, pollutants, UV radiation, temperature changes, and other harmful substances. To maintain their quality and properties, various methods and means of surface protection are employed, tailored to specific requirements and operational conditions.

Figure 6.
Hydrophobic coating effect [28].

One of the pioneering studies in this field is the work of Dario S. Facio and Maria J. Mosquera [28], in which the researchers introduced a simple strategy for producing superhydrophobic nanocomposite coatings in situ on a construction substrate. Their method, based on an innovative approach to nanocomposites, opens up new possibilities for creating durable and efficient hydrophobic coatings for various construction applications. **Figure 6** illustrates the hydrophobic effect of the material without and with the coating.

In the context of advancements in superhydrophobic surfaces, the work of Yun-Yun Quan et al. [31] provides a review of the latest achievements in creating durable superhydrophobic surfaces. The authors focus on various aspects of structures and materials that contribute to achieving desired hydrophobic properties, emphasizing the evolution of techniques and materials used in this field.

Another significant contribution in the realm of hydrophobic coatings is the study conducted by Zhang et al. [32]. The researchers presented functional and versatile superhydrophobic coatings obtained through stoichiometric silanization. Their approach to creating hydrophobic coatings, which are both durable and versatile, holds potential for application across a broad spectrum of industries, from construction to manufacturing.

In summary, the development of hydrophobic coating technologies in recent years has brought about many innovative solutions that have the potential to revolutionize the construction sector and other fields. Thanks to studies like those presented by Facio, Mosquera, Quan, Chen, and Zhang, the future of hydrophobic coatings seems promising, offering materials with improved performance, durability, and functionality.

4.2.1 Painting and impregnation

One of the most popular and effective approaches to modifying the wetting of building materials is through surface coating techniques. The use of paints and impregnating agents allows the creation of a protective layer on the material's surface, shielding it from water infiltration and other substances. The coating can also serve esthetic purposes, providing the desired color and appearance to the material's Surface [33]. By applying a hydrophobic material layer to the surface, the material's interaction with liquids can be significantly altered. The study conducted by Bing Yin et al.

[34], for example, focused on applying superhydrophobic coatings to concrete surfaces. Their research indicates that such coatings effectively reduce water absorption by concrete, potentially enhancing its durability in moist conditions. This example underscores the role of hydrophobic coatings in improving the ability of building materials to repel water.

In the study conducted by Yao et al. [9], the authors thoroughly analyzed the advancements in hydrophobic cement-based materials. Their research focused on the preparation, characterization, and properties of these materials, emphasizing the importance of appropriate coating techniques to achieve desired hydrophobic properties. This study underscores that proper painting and impregnation techniques can significantly influence the ability of building materials to repel water, which is crucial for maintaining their integrity and extending their lifespan in various environmental conditions.

In a comprehensive review by Pan et al. [35], the authors delved into various methods of concrete surface treatment. Their research, encompassing a wide range of techniques, highlighted the significance of painting and impregnation in the realm of building materials. The study meticulously categorizes different types of treatments and elucidates their underlying mechanisms. This work underscores the pivotal role that surface treatments play in enhancing the durability, performance, and esthetic appeal of concrete structures. By understanding the intricacies of these methods, professionals can make informed decisions to ensure the longevity and resilience of their constructions in the face of environmental challenges.

Surface coating techniques serve as versatile tools for modifying the wetting properties of building materials. The research examples illustrate that the application of protective coatings can significantly impact the durability, performance, and longevity of materials in humid and adverse conditions.

4.2.2 Hydrophobic and oleophobic coatings

Hydrophobic and oleophobic coatings have gained widespread attention due to their exceptional ability to provide protection against water and oil. These coatings work by altering the surface properties of materials, making them less prone to wetting and absorption while promoting fluid runoff instead of absorption. This inherent feature makes these coatings particularly effective in safeguarding materials constantly exposed to moisture, including building facades, roofs, and architectural elements (**Figure** 7) [36–48].

Coating the surface of cement-based materials with substances of low surface free energy, such as silanes and fluorosilicates [37]. The primary mechanism of this process is the reaction of the molecular functional groups of these substances with the hydroxyl groups on the surface of cementitious materials. As a result, a layer of hydrophobic molecules forms on the material's surface, converting it into a hydrophilic one [38, 39].

The differences between silane-based hydrophobic coatings and traditional waterproof coatings mainly arise from the molecular characteristics of silanes. Due to their small molecular structure, silanes can easily penetrate open cracks or capillary pores in the material, creating an effective hydrophobic barrier [40].

Contemporary research in the field of hydrophobic modification increasingly combines various concepts and techniques. For instance, the studies by Genga et al. focused on combining surface roughness with graphene oxide and reducing surface

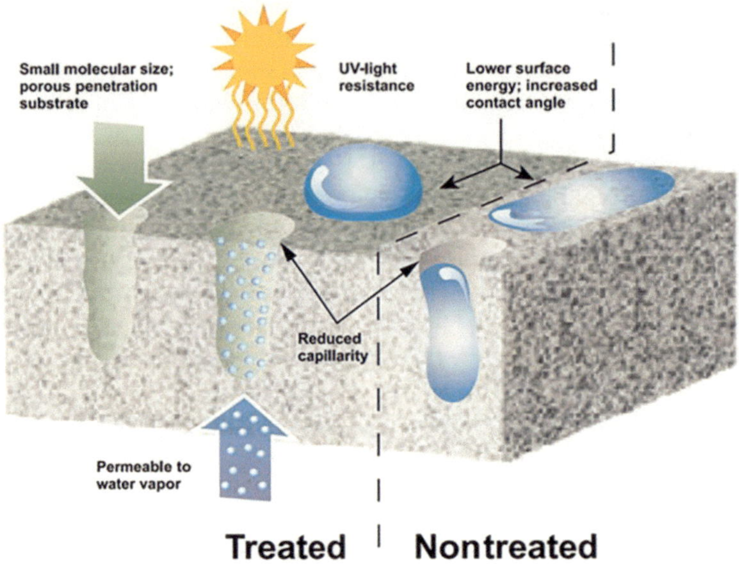

Figure 7.
Silicone-based products penetrate deeply, forming a repellent layer within the substrate [36].

energy using a silane solution [41]. Other research, such as that conducted by Wong et al., centered on utilizing waste in the process of creating hydrophobic coatings [42].

An important aspect in the context of hydrophobic coatings is also the use of nanoparticles. Their presence can significantly increase surface roughness, leading to enhanced hydrophobic properties [43–45]. Nanoparticles can also influence the structure and compactness of coatings, further increasing their resistance to water penetration [46, 47].

In the context of preserving cultural heritage, protecting building materials from degradation is a pivotal issue. In response to this challenge, Mosquera, Carrascosa, and Badreldin introduced an innovative approach to creating superhydrophobic and oleophobic coatings on historic building materials [48]. Their research focuses on techniques and methods that effectively enhance not only the resistance of materials to water and oils but are also suitable for delicate and historical surfaces (**Figure 8**). Through this approach, it is possible to maintain the integrity and authenticity of

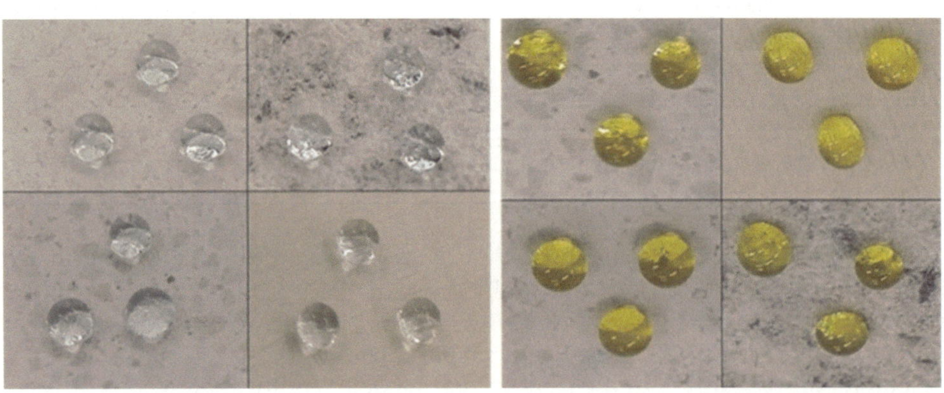

Figure 8.
The effect of hydro- and oleophobic coatings [48].

historic structures while providing them protection against damaging factors. This work represents a significant step toward integrating modern technologies with the conservation of cultural heritage.

The effectiveness of hydrophobic and oleophobic coatings in safeguarding building materials, surfaces, and architectural elements is highlighted by the presented research. By altering surface properties to deter water and oil interactions, these coatings play a crucial role in enhancing the resistance and esthetic preservation of structures in various environmental conditions.

4.2.3 Anti-graffiti agents

In urban settings where graffiti poses a frequent issue, the application of anti-graffiti agents has become a common solution aimed at protecting surfaces from permanent graffiti painting. In **Figure 9**, a schematic of the anti-graffiti coating is presented. These agents play a crucial role in creating a protective layer, which in turn facilitates the removal of unwanted paints and marks while minimizing damage to the substrate.

Modern cities around the world grapple with the issue of graffiti, which, although often considered a form of street art, can lead to the degradation and damage of building materials. In response to this challenge, Amrutkar et al. [49] conducted an in-depth analysis of the latest solutions in the field of anti-graffiti coatings.

In their study, the authors focused on various technologies and materials used in anti-graffiti coatings. They highlighted the fact that an ideal anti-graffiti coating should not only effectively protect the surface from spray paint but also be durable, weather-resistant, and easy to maintain.

One of the key findings of the study was that anti-graffiti coatings can be divided into two main categories: permanent and removable. Permanent coatings create a durable barrier on the surface of the building material, preventing paint adhesion. On

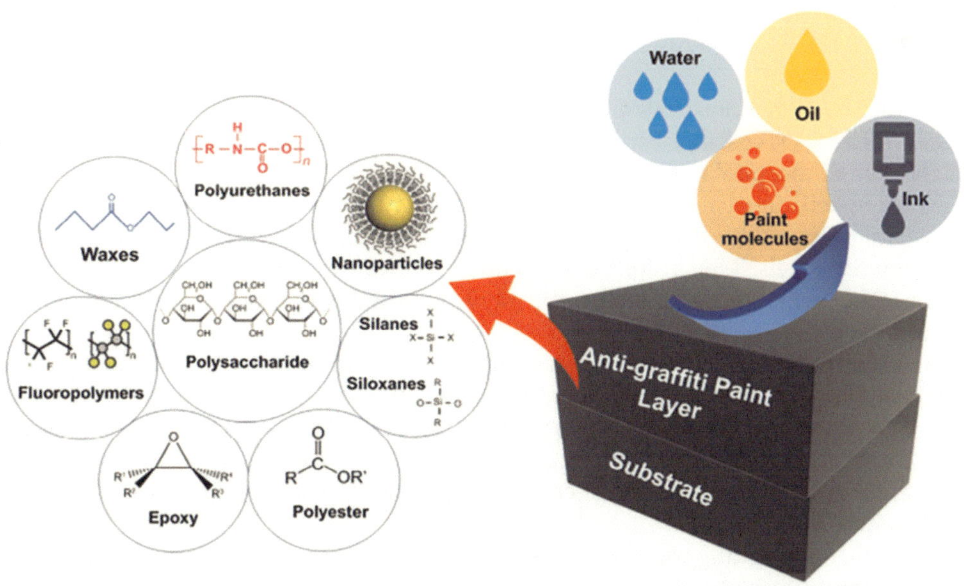

Figure 9.
Anti-graffiti coating [49].

the other hand, removable coatings allow for easy removal of graffiti without damaging the surface.

The authors [49] also emphasized the importance of nanotechnology in creating modern anti-graffiti coatings. Thanks to the use of nanoparticles and nanocomposites, it is possible to create coatings with exceptional resistance to various types of paints and markers, while ensuring excellent adhesion to various building materials.

In the context of sustainable development and ecology, attention should be paid to the need to create environmentally friendly anti-graffiti coatings. The publication [49] mentioned research on biodegradable and ecological coatings, which effectively protect not only against graffiti but also minimize the negative impact on the environment.

In conclusion, the review publication [49] provides a comprehensive overview of the field of anti-graffiti coatings, highlighting the latest achievements and directions of development. Thanks to such studies, it is possible to create effective and durable solutions that help maintain the esthetics and durability of urban structures in the face of challenges associated with graffiti.

The application of anti-graffiti agents presents a valuable strategy for protecting urban surfaces from permanent graffiti. Research underscores the efficacy of these coatings in deterring vandalism, facilitating maintenance, and preserving cultural heritage. In cities grappling with graffiti-related challenges, the use of anti-graffiti agents remains a crucial component of conservation and urban environment maintenance.

4.2.4 Self-cleaning coatings

The urban environment is constantly exposed to various pollutants, which can degrade the appearance and functionality of building materials. In recent years, the development of self-cleaning coatings has emerged as a promising solution to address this challenge. These coatings maintain not only the esthetics of structures but also contribute to their longevity by reducing the need for frequent cleaning and maintenance.

In a study by Cedillo-González et al. [50], the authors delved into the self-cleaning properties of TiO_2 coatings for building materials. Their research highlighted the crucial role of morphology and humidity in the stain removal performance of these coatings. Specifically, they found that the efficiency of TiO_2 coatings in removing stains is significantly influenced by their morphological characteristics and the ambient humidity levels. This insight underscores the importance of tailoring the properties of self-cleaning coatings to specific environmental conditions to achieve optimal performance.

Rabajczyk et al. [51] took a broader perspective, examining self-cleaning coatings and surfaces of modern building materials designed to remove various air pollutants. Their research emphasized the potential of these coatings to keep surfaces not only clean but also contribute to improving air quality. By effectively removing pollutants from their surfaces, these coatings can play a pivotal role in creating healthier urban environments.

One of the intriguing aspects of self-cleaning surfaces is their ability to interact with different types of pollutants. Quan et al. [52] explored the role of surface wettability and dust types in the self-cleaning process. Their findings revealed that the efficiency of self-cleaning is significantly influenced by the type of dust and the wettability of the surface. For instance, certain dust types can be more easily removed from hydrophobic surfaces, while others may require hydrophilic properties for effective cleaning.

The mechanism behind many self-cleaning coatings is the photocatalytic activity of materials like TiO_2. When exposed to sunlight, these materials can break down organic pollutants, making it easier for them to be washed away by rain or wind. This not only ensures cleaner surfaces but also reduces the organic load on building materials, potentially extending their lifespan [50].

Another significant advantage of self-cleaning coatings is their potential economic benefits. By reducing the frequency and intensity of cleaning required, these coatings can lead to significant savings in maintenance costs over the lifespan of a building or structure [51].

However, it is essential to note that while self-cleaning coatings offer numerous benefits, they are not a one-size-fits-all solution. As highlighted by the studies, the effectiveness of these coatings can vary based on environmental conditions, the specific pollutants in question, and the properties of the coating itself [52].

In conclusion, the development and application of self-cleaning coatings represent a significant advancement in the field of building materials. As urban environments continue to grapple with pollution and its associated challenges, these coatings offer a promising avenue to maintain the esthetics, functionality, and longevity of structures. Self-cleaning coatings represent an innovative solution in the field of building materials and are crucial for maintaining the esthetics, durability, and functionality of surfaces. Research examples on the applications of self-cleaning coatings for building materials highlight the diversity of applications and the potential of such solutions. As the construction industry emphasizes durability, performance, and esthetics, self-cleaning coatings remain a key tool in achieving these goals.

5. Application of hydrophobic coatings in practice

To test the surface protection capabilities of concrete, geopolymers, and sandstone, 19 commercially available products were tested. These products were selected based on their specific applications, particularly those intended for surface treatment of concrete or other mineral materials, as well as in accordance with previous research. Furthermore, the selected products were primarily designed for surface hydrophobization, such as primer coatings (mainly used as an adhesive bridge between the surface and other coatings, but often applied independently for surface hydrophobization).

Below are the coatings that exhibited high hydrophobicity:

Stoprim micro - A primer, transparent coating for mineral substrates based on a concentrated aqueous solution of siloxanes. Depending on the dilution ratio with water, it can be used as a primer or for hydrophobization. For the following tests, a dilution ratio of 1:4 was used.

Lukofob classic - Hydrophobic coating based on polysiloxane resin in organic solvents. Intended for treatment of mineral surfaces (natural stones, concrete, etc.). To be used in an undiluted form.

***Repesil* Aqua** - Hydrophobic coating based on a water emulsion of siloxanes. Designed for absorbent mineral substrates (discolorations may occur on weakly absorbent substrates). To be used in an undiluted form.

Hydrofobizant Fortesil - Final hydrophobic coating for natural and artificial stone, based on a water emulsion of siloxanes. To be used in an undiluted form.

To determine the coating's ability to hydrophobize surfaces, a series of tests were performed, including the drop test described in the earlier part of the chapter, and the results are presented below.

5.1 Stoprim micro

This product achieved very good results in hydrophobizing the surfaces of the tested materials. The initial contact angles were 125.1° for concrete, 130.5° for geopolymers, and 119.5° for sandstone. At the same time, the droplet size changed only slightly during measurements after several minutes, indicating more of an evaporation process than infiltration. Furthermore, the appearance of the samples remained unchanged (**Table 1**).

5.2 Lukofob classic

This product showed excellent results in hydrophobization. The initial contact angle was 137° for concrete, 133.8° for geopolymers, and 129.8° for sandstone. At the same time, the droplet size changed only slightly during measurements after several minutes, indicating more of an evaporation process than infiltration. The droplet size also changed only slightly. The appearance of the samples remained unchanged after application (**Table 2**).

t(s)	Concrete	Geopolymer	Sandstone
		Contact angle (°)	
0	125.1	130.5	119.5
1	120.1	128.7	118.6
2	119.8	128.2	117.0
3	118.5	125.1	116.5
4	117.0	125.0	114.3
5	115.8	123.6	111.9

Table 1.
Contact angles of samples subjected to the action of Stoprim micro.

t(s)	Concrete	Geopolymer	Sandstone
		Contact angle (°)	
0	137.0	133.8	129.8
1	134.4	131.7	128.0
2	133.4	130.0	127.9
3	133.0	129.5	127.5
4	132.7	127.8	126.6
5	132.3	127.0	126.0

Table 2.
Contact angles of samples subjected to the action of Lukofob classic.

5.3 Repesil aqua

This product showed excellent results in hydrophobization of the tested materials' surfaces. The initial contact angle was 131.6° for concrete, 124.2° for geopolymers, and 121.9° for sandstone. The droplet size varied only slightly, and the appearance of the samples remained unchanged after application. The main advantage of this product is that it is water-based (water emulsion of siloxanes), which reduces the risk of contamination or damage to treated objects by potentially aggressive solvents (**Table 3**).

5.4 Hydrofobizator fortesil

Although this product did not alter the appearance of the samples and effectively hydrophobized sandstone and concrete, achieving high contact angles and minimal droplet changes, it did not perform as well on geopolymers. While the initial contact angle was very high for geopolymers, the droplet quickly infiltrated, and the infiltration rate was highly inconsistent. This suggests that although the agent is strongly hydrophobic, it is unable to fill the highly porous surface of geopolymers (**Table 4**).

t(s)	Concrete	Geopolymer	Sandstone
		Contact angle (°)	
0	131.6	124.2	121.9
1	125.0	120.8	119.9
2	124.7	117.3	118.3
3	123.2	114.7	116.2
4	121.0	114.3	115.9
5	120.2	112.4	113.9

Table 3.
Contact angles of samples subjected to the action of Repesil aqua.

t(s)	Concrete	Geopolymer	Sandstone
		Contact angle (°)	
0	129.0	114.9	129.0
1	126.5	103.4	127.3
2	125.0	84.2	125.9
3	123.8	45.1	125.1
4	122.0	—	124.0
5	120.9	—	119.6

Table 4.
Contact angles of samples subjected to the action of hydrophobic agent Fortesil.

6. Conclusions

The chapter "Hydrophobic Coatings for Building Materials" explores the significance of modifying wetting properties to enhance the durability and performance of construction materials. It presents research conducted on commonly used building materials such as Portland cement and sandstone, as well as new materials like geopolymers. To introduce readers to the realm of construction materials, the first part of the chapter compares alternative materials, such as geopolymers, with the widely used Portland cement, highlighting their distinct reactions during the solidification process, as well as their physicochemical, mechanical properties, and environmental impact.

The concept of wetting is comprehensively discussed in the context of building materials, with a particular focus on factors influencing wetting behavior, such as pore size and distribution, chemical composition, and surface properties. Various methods for assessing wetting are presented, including the spreading method, laser beam method, and drop method.

The chapter underscores the importance of surface protection for building materials. In its subsequent sections, it elaborates on the significance of modifying wetting behavior through bulk additives and coatings, providing examples of nanomaterials and polymers employed to alter surface properties. Strategies such as hydrophobic and oleophobic coatings and self-cleaning coatings are discussed, aiming to enhance the durability and performance of materials in diverse environments.

The final part of the chapter presents research on selected hydrophobic coatings aimed at extending the lifespan of building constructions. During tests of hydrophobic protection, a range of commercially available products dedicated to protecting concrete and other mineral materials were examined, evaluating their impact on water absorption on the surfaces of concrete, geopolymers, and sandstone (and additionally sandstone) using the drop test method. The testing of hydrophobization on these materials revealed that many of the products already available on the market can also be applied to protect the surfaces of geopolymers and sandstone. The most effective were found to be hydrophobic coatings based on siloxanes and protective coatings based on epoxy resins.

This chapter provides fundamental knowledge about hydrophobic protection for building materials and serves as a basis for further research, demonstrating their efficacy in modifying surface wetting properties and enhancing material protection.

Acknowledgements

This chapter was supported by the project titled "Development of Geopolymer Composites as a Material for the Protection of Hazardous Wrecks and Other Critical Underwater Structures Against Corrosion," with project number TH80020007. The support was provided by the Technology Agency of the Czech Republic (TACR) under the Epsilon Program during the 2021 M-ERA.Net3 Call. Additionally, the chapter was carried out at the Technical University of Liberec, Faculty of Mechanical Engineering, with the Institutional Endowment for the Long-Term Conceptual Development of Research Institutes, as provided by the Ministry of Education, Youth, and Sports of the Czech Republic in 2023.

Author details

Katarzyna Buczkowska
Faculty of Mechanical Engineering, Technical University of Liberec, Czech Republic

*Address all correspondence to: katarzyna.ewa.buczkowska@tul.cz

References

[1] Zhao J, Gao X, Chen S, Lin H, Li Z, Lin X. Hydrophobic or superhydrophobic modification of cement-based materials: A systematic review. Composites Part B: Engineering. 2022;**243**(15):110104. DOI: 10.1016/j.compositesb.2022.110104

[2] Neville AM. Properties of Concrete. 5th ed. Pearson; ISBN 978-0-273-75580-7

[3] Hardjito D, Wallah SE, Sumajouw DMJ, Rangan BV. On the development of fly ash-based geopolymer concrete. ACI Materials Journal. 2004;**101**(6):467-472

[4] Provis JL, van Deventer JSJ. Geopolymers, Structures, Processing, Properties and Industrial Applications. Woodhead Publishing Series in Civil and Structural Engineering; 2009. pp. 1-11

[5] Kumar Mehta P, Monteiro PJM. Concrete: Microstructure, Properties, and Materials. 4th ed. McGraw-Hill Education; 2014 ISBN: 9780071797870

[6] Mindess S, Young JF, Darwin D. Concrete. 2nd ed. Upper Saddle River: Prentice-Hall; 2003

[7] Davidovits J. Geopolymer Chemistry and Applications. 5th ed. Saint-Quentin: Geopolymer Institute; 2020 ISBN: 9782954453118

[8] Václav G. Geopolymery: výroba, vlastnosti a použití [thesis]. Ostrava: Bakalářská práce. Vysoká Škola Báňská - Technická Univerzita Ostrava; 2013

[9] Yao H, Xie Z, Huang C, Yuan Q, Zhiwu Y. Recent progress of hydrophobic cement-based materials: Preparation, characterization and properties. Construction and Building Materials. 2021;**299**(13):124255. DOI: 10.1016/j.conbuildmat.2021.124255

[10] Liu J, Song L, Cui C, Liu R, Zhiwu Y. Drying and wetting mechanisms of cementitious materials under natural climate: Models, simulations, and experiments. Construction and Building Materials. 2022;**338**:127532, ISSN 0950-0618. DOI: 10.1016/j.conbuildmat.2022.127532

[11] Coudert F-X, Boutin A, Fuchs AH. Open questions on water confined in nanoporous materials. Communications Chemistry. 2021;**4**:106. DOI: 10.1038/s42004-021-00544-9

[12] Li C, Jiang L. Effect of limestone powder addition on corrosion initiation time of reinforced concrete. Journal of Building Engineering. 2022;**59**:105132. DOI: 10.1016/j.jobe.2022.105132

[13] Zhao H, Ding J, Huang Y, Tang Y, Wen X, Huang D. Experimental analysis on the relationship between pore structure and capillary water absorption characteristics of cement-based materials. Structural Concrrete. 2019;**20**(5):1750-1762. DOI: 10.1002/suco.201900184

[14] Wang X, Fu C, Zhang C, Qiu Z, Wang BA. Comprehensive review of wetting transition mechanism on the surfaces of microstructures from theory and testing methods. Materials. 2022;**15**:4747. DOI: 10.3390/ma15144747

[15] Xu Z, Masliyah JH. Contact angle measurement on oxide and related surfaces. In: Hubbard A, editor. Encyclopedia of Surface and Colloid Science. New York: Marcel Dekker; 2002. pp. 1228-1241

[16] Shirong Zhu, Kaibin Xie, Qiaoli Lin, Rui Cao, Feng Qiu. Experimental

determination of surface energy for high-energy surface: A review, Advances in Colloid and Interface Science Volume 315, May 2023, 102905, doi: 10.1016/j.cis.2023.102905

[17] Ding F, Gao M. Pore wettability for enhanced oil recovery, contaminant adsorption and oil/water separation: A review. Advances in Colloid and Interface Science. 2021;**289**:102377. DOI: 10.1016/j.cis.2021.102377

[18] Grzegorczyk-Frańczak M, Barnat-Hunek D, Łagód G, Pavlík Z. Changes of wetting properties and surface free energy at the time of hydrophobized concretes with boiler slag and coal combustion dust. AIP Conference Proceedings. 2020;**2305**:020005. DOI: 10.1063/5.0034844

[19] Adhikary SK, Ashish DK, Sharma H, Patel J, Rudzionis Z, Al-Ajamee M, et al. Lightweight self-compacting concrete: A review. Resources, Conservation & Recycling Advances. 2022;**15**:200107. DOI: 10.1016/j.rcradv.2022.200107

[20] Xiao J, Lv Z, Duan Z, Zhang C. Pore structure characteristics, modulation and its effect on concrete properties: A review. Construction and Building Materials. 2023;**397**(15):132430. DOI: 10.1016/j.conbuildmat.2023.132430

[21] Jin S, Zhou J, Xiyao Zhao L, Sun. Quantitative relationship between pore size distribution and compressive strength of cementitious materials. Construction and Building Materials. 2021;**273**(1):121727. DOI: 10.1016/j.conbuildmat.2020.121727

[22] Zou D, Li K, Li W, Li H, Cao T. Effects of pore structure and water absorption on internal curing efficiency of porous aggregates. Construction and Building Materials. 2018;**163**(28): 949-959. DOI: 10.1016/j.conbuildmat.2017.12.170

[23] Chen S, Ruan S, Zeng Q, Yi L, Mingzhong Z, Tian Y, et al. Pore structure of geopolymer materials and its correlations to engineering properties: A review. Construction and Building Materials. 2022;**328**(18):127064. DOI: 10.1016/j.conbuildmat.2022.127064

[24] Bang J, Choi J, Hong W-T, Jung J, Kim GM, Yang B. Influences of binder composition and carbonation curing condition on the compressive strength of alkali-activated cementitious materials: A review. Journal of CO_2 Utilization. 2023;**74**:102551. DOI: 10.1016/j.jcou.2023.102551

[25] Ruan Y, Jamil T, Hu C, Gautam BP, Yu J. Microstructure and mechanical properties of sustainable cementitious materials with ultra-high substitution level of calcined clay and limestone powder. Construction and Building Materials. 2022;**314**(Part A):125416. DOI: 10.1016/j.conbuildmat.2021.125416

[26] Anovitz LM, Cole DR. Characterization and analysis of porosity and pore structures. Reviews in Mineralogy and Geochemistry. 2015; **80**(1):61-164. DOI: 10.2138/rmg.2015.80.04

[27] Mora E, González G, Romero P, Castellón E. Control of water absorption in concrete materials by modification with hybrid hydrophobic silica particles. Construction and Building Materials. 2019;**221**:210-218. DOI: 10.1016/j.conbuildmat.2019.06.086

[28] Facio DS, Mosquera MJ. Simple strategy for producing superhydrophobic nanocomposite coatings in situ on a building substrate. ACS Applied Materials & Interfaces.

2013;**5**(15):7517-7526. DOI: 10.1021/am401826g

[29] Li S, Sng A, Daniel D, Lau HC, Torsæter O, Stubbs LP. Visualizing and quantifying wettability alteration by silica nanofluids. ACS Applied Materials & Interfaces. 2021;**13**(34):41182-41189. DOI: 10.1021/acsami.1c08445

[30] Sicakova A, Draganowska M, Kovac M. Characteristic of building mixes containing fine particles of selected recycled materials. Procedia Engineering. 2017;**180**:1256-1265. DOI: 10.1016/j.proeng.2017.04.287

[31] Quan Y-Y, Chen G, Lai Y, Huang Z-S, Li H. Recent advances in fabricating durable superhydrophobic surfaces: A review in the aspects of structures and materials. Materials Chemistry Frontiers. 2021;**5**:1655-1682. DOI: 10.1039/D0QM00485E

[32] Zhang L, Zhou AG, Sun BR, et al. Functional and versatile superhydrophobic coatings via stoichiometric silanization. Nature Communications. 2021;**12**:982. DOI: 10.1038/s41467-021-21219-y

[33] Otitoju TA, Ahmad AL, Ooi BS. Superhydrophilic (superwetting) surfaces: A review on fabrication and application. Journal of Industrial and Engineering Chemistry. 2017;**47**:19-40, ISSN 1226-086X,. DOI: 10.1016/j.jiec.2016.12.016

[34] Yin B, Tianyuan X, Hou D, Zhao E, Hua X, Han K, et al. Superhydrophobic anticorrosive coating for concrete through in-situ bionic induction and gradient mineralization. Construction and Building Materials. 2020;**257**:119510, ISSN 0950-0618. DOI: 10.1016/j.conbuildmat.2020.119510

[35] Pan X, Shi Z, Shi C, Ling T-C, Li N. A review on concrete surface treatment part I: Types and mechanisms. Construction and Building Materials. 2017;**132**:578-590, ISSN 0950-0618,. DOI: 10.1016/j.conbuildmat.2016.12.025

[36] Construction Chemical Solutions for Building Materials Protection Real solutions for real market needs [Internet]. 2018. Available from: https://www.dow.com/content/dam/dcc/documents/en-us/catalog-selguide/63/63-10/63-1065-01-building-materials-protection-chemical-solutions.pdf [Accessed: September 2, 2023]

[37] Chen D, Hong C, Zhang M, She AY, Qiao ZF, Wang YM. Mechanism and study method of silane impregnated concreto. China Build Waterproofing. 2015;**9**:17-21

[38] Woo RSC, Zhu HG, Chow MMK, Leung CKY, Kim JK. Barrier performance of silane–clay nanocomposite coatings on concrete structure. Composites Science and Technology. 2008;**68**(14):2828-2836

[39] Tian TW, Ni Y, Liu GB, Wang YD, Li P. Silane impregnation for protection of concrete bridges. China Building Waterproofing. 2011;**12**:4-7

[40] Ran L, Li X, Jiang L. Coastal hydraulic concrete structure surface waterproof technology. Science, Technology and Engineering. 2019; **19**(23):222-227

[41] Geng YJ, Li SC, Hou DS, Zhang WF, Jin ZQ, Li QY, et al. Fabrication of superhydrophobicity on foamed concrete surface by GO/silane coating. Materials Letters. 2020;**265**:127423

[42] Wong HS, Barakat R, Alhilali A, Saleh M, Cheeseman CR. Hydrophobic concrete using waste paper sludge ash. Cement and Concrete Research. 2015;**70**:9-20

[43] Manoudis PN, Karapanagiotis I. Modification of the wettability of polymer surfaces using nanoparticles. Progress in Organic Coatings. 2014;**77**: 331-338

[44] Yüce MY, Demirel AL, Menzel F. Tuning the surface hydrophobicity of polymer/nanoparticle composite films in the Wenzel regime by composition. Langmuir: ACS Journal of Surfaces and Colloids. 2005;**21**:5073-5078

[45] Manoudis PN, Karapanagiotis I, Tsakalof A, Zuburtikudis I, Panayiotou C. Superhydrophobic composite films produced on various substrates. Langmuir. 2008;**24**: 11225-11232

[46] Hou PK, Li R, Li QF, Lu N, Wang ML, Cheng X, et al. Novel superhydrophobic cement-based materials achieved by construction of hierarchical surface structure with FAS/ SiO_2 hybrid nanocomposites. ES Materials & Manufacturing. 2018;**1**:57-66

[47] Li G, Yue J, Guo CH, Ji YS. Chloride resistance of concrete containing nanoparticle-modified polymer cement-based coatings. Construction and Building Materials. 2017;**299**:123736

[48] Mosquera MJ, Carrascosa LAM, Badreldin N. Producing superhydrophobic/oleophobic coatings on cultural heritage building materials. Pure and Applied Chemistry. 2018;**90**(3): 551-561. DOI: 10.1515/pac-2017-0404

[49] Amrutkar S, More A, Mestry S, et al. Recent developments in the anti-graffiti coatings: An attentive review. Journal of Coating Technology and Research. 2022; **19**:717-739. DOI: 10.1007/s11998-021-00580-z

[50] Cedillo-González EI, Hernández-López JM, Ruiz-Valdés JJ, Barbieri V, Siligardi C. Self-cleaning TiO_2 coatings for building materials: The influence of morphology and humidity in the stain removal performance. Construction and Building Materials. 2020;**237**:117692, ISSN 0950-0618. DOI: 10.1016/j.conbuildmat.2019.117692

[51] Rabajczyk A, Zielecka M, Klapsa W, Dziechciarz A. Self-cleaning coatings and surfaces of modern building materials for the removal of some air pollutants. Materials. 2021;**14**:2161. DOI: 10.3390/ma14092161

[52] Quan Y-Y, Zhang L-Z, Qi R-H, Cai R-R. Self-cleaning of surfaces: The role of surface wettability and dust types. Scientific Reports;**6**:38239. DOI: 10.1038/srep38239